COMBINATION OF OBSERVATIONS

BY THE SAME AUTHOR

Spherical Astronomy
Stellar Dynamics
Celestial Mechanics
The Origin of the Earth
Some Famous Stars
Foundations of Astronomy
Astronomy
The Sun, the Stars and the Universe
Astrophysics
John Couch Adams and the Discovery of Neptune
Foundations of Analytical Geometry
Sea and Air Navigation
Introduction to Sea and Air Navigation
Astronomical Navigation
Handbook of Sea Navigation

ALSO JOINT-AUTHOR OF

Admiralty Manual of Navigation (1922), Vols. 1, 2
Position Line Tables

COMBINATION OF OBSERVATIONS

BY

W. M. SMART, M.A., D.Sc.

Regius Professor of Astronomy in the University of Glasgow

CAMBRIDGE

AT THE UNIVERSITY PRESS

1958

CAMBRIDGE UNIVERSITY PRESS
Cambridge, New York, Melbourne, Madrid, Cape Town, Singapore, São Paulo, Delhi

Cambridge University Press
The Edinburgh Building, Cambridge CB2 8RU, UK

Published in the United States of America by Cambridge University Press, New York

www.cambridge.org
Information on this title: www.cambridge.org/9780521064903

First published 1958
This digitally printed version 2008

A catalogue record for this publication is available from the British Library

ISBN 978-0-521-06490-3 hardback
ISBN 978-0-521-09609-6 paperback

CONTENTS

Chapter 6: EQUATIONS OF CONDITION IN SEVERAL UNKNOWNS

Chapter 7: THEORETICAL FREQUENCY DISTRIBUTIONS

Chapter 8: CORRECTION OF STATISTICS

PREFACE

The subject-matter of this volume may be summarised briefly as the study of frequency distributions encountered in many branches of experimental science, and the treatment of observations and measures in which the incidence of errors is recognised. Most topics in the book have formed the substance of lectures for students proceeding to an Honours Degree in the University of Glasgow, during the past two decades.

Chapter 1 deals with frequency distributions as viewed against the background of general statistical theory; the various techniques relating to moments are considered in some detail, together with the integrals, associated with the normal function, which form the basis for many of the subsequent investigations.

The next five chapters are concerned with the treatment of observations and measures subject to errors, the general foundation being the normal law of errors. It is not too much to affirm that an observational or experimental result is to be judged in the absence of systematic error, not by its apparent agreement or disagreement with some other result previously obtained, but on the ascertainable degree of precision of the actual observations or measures relating to the investigation concerned. It is then necessary that the rules for deriving the degree of precision of a particular result should be expressed in some standard form which has universal sanction. In much of the earlier literature one of the measures of precision is known by the cumbrous expression of 'root-mean-square error'; I have replaced this important quantity by the simpler expression of 'standard error' which brings it into line with its counter-part, 'standard deviation', in the theory of statistics. Proofs of the normal law of errors and associated theorems are given in these chapters, with due regard to the various hypotheses which it is necessary to introduce if a mathematical formulation is to be achieved.

In several examples worked out in the text, one being intentionally of a highly artificial character, my aim has been to introduce arithmetic simplicity so that attention might not be diverted, by elaborate computational details, from fundamental principles and practical methods of procedure; the complete investigation of Gauss's well-known example involving three unknowns (Chapter 6) is a case in point.

Chapter 7 deals with the representation of a frequency distribution in various ways according to the nature of the series of measures recorded in particular observations or experiments, the object being to derive a mathematical expression which best represents the practical results; in particular, there are detailed discussions of the representation of measures by a Gram-Charlier series and of distributions derivable from Pearson's differential equation, with an example of the latter taken from current physical literature.

Chapter 8 deals with the correction of frequency distributions, the subject-matter including Sheppard's corrections to moments, the application of Fourier transforms and the correction of vectors.

The final chapter introduces the subject of correlation, the use of contingency tables, and a discussion of normal bivariate and trivariate distributions.

In the appendices are to be found tables of several functions including, in particular, the error function; as regards the latter I am indebted to the Royal Society of Edinburgh for permission to make use of the elaborate tables calculated by J. Burgess (*Trans. Roy. Soc. Edinb.* **39** (1898), 257–321).

It may be added that no elaborate equipment is necessary for the statistical conduct of the various types of investigation with which the book deals; *Barlow's Tables, Crelle's Rechentafeln* (multiplication tables) the usual collection of logarithmic tables, and possibly a calculating machine, suffice for the arithmetic computations.

In conclusion I am again much indebted to the officials and staff of the Cambridge University Press for their care and attention during the course of printing.

W.M.S.

UNIVERSITY OBSERVATORY
GLASGOW

15 *June 1957*

CHAPTER 1

FREQUENCY DISTRIBUTIONS

1·01. Introduction

In this chapter we deal with some general principles from the viewpoint of the theory of statistics in preparation, to some extent at least, for the more specialized study of the theory of errors and related subjects. Statistics is a branch of mathematics concerned, in its simplest stages, with the study of the detailed information relating to some specific characteristic such as the weight measures of national service recruits or the marks scored in an examination; in some of the subsequent chapters the emphasis is on the study of observations or measures in which the appearance of unavoidable errors is fully recognized.

As a typical example, illustrating the initial methods of statistics, suppose that the school authorities of a city provide the individual measures of the heights of a thousand boys belonging to a particular age group. This mass of information is not readily comprehended as it stands, and the first task of the statistician is to have the information arranged in a suitably manageable or condensed form. This is done most conveniently by deriving the numbers of boys with heights, for example, between 40 and 41 in., between 41 and 42 in., and so on; there is now a much clearer picture of the numerical distribution according to height groups. This condensed information can now be displayed by means of a graphical representation from which the distribution in the several groups can be seen at a glance.

Associated with this distribution are certain calculable quantities which specify the chief characteristics of the distribution (with these we deal later), and so the original mass of information relating to the heights of a thousand boys, perhaps arranged initially in the most haphazard fashion, is finally replaced by the graphical representation and these calculable quantities. It is thus possible to interpret the statistics as a whole and, possibly, to suggest uses to which this information can be applied.

A second example of a less simple nature, taken from astronomy, concerns the stars of the *main sequence* for which detailed information had been accumulated relating, in particular, to luminosity (or intrinsic brightness), effective temperature (or, more crudely, surface temperature) and mass. In the early years of the present century it was found that there existed a definite relationship between luminosity and spectral type, later translated in terms of temperature. Further, it was seen that the most luminous stars of the main sequence were also the most massive, and the feeblest the least massive. Here was a challenge to the theoretical investigator of the physical conditions within a star, culminating (in 1924) in Eddington's discovery† of the 'mass-luminosity relationship'.

Not infrequently, the characteristics of a statistical distribution relating to a series of observations or measures have led to a greater insight into the particular problem under consideration and to new discoveries such as Bradley's discovery of aberration and nutation referred to in §2·04 (p. 39).

1·02. Frequency distributions

We consider as an example the statistics relating to the heights of a thousand men in a regiment. To condense the information conveniently, we derive the number of men in each of seven groups according to height; the first group relates to heights between 62 and 64 in., with 63 in. as the 'middle height'; the second group relates to heights between 64 and 66 in., with 65 in. as the middle height; and so on.

The number in any group is called the *frequency* for that group, usually denoted in statistical theory by f, with suffices 1, 2, ... to indicate the group concerned;‡ thus the frequency for the ith group is denoted by f_i. The statistics for the seven groups are shown in Table 1, the middle heights being shown in the second row and the corresponding frequencies in the third row.

Table 1. *Distribution of heights of* 1000 *soldiers*

Group ...	1	2	3	4	5	6	7
Middle height (in.)	63	65	67	69	71	73	75
Frequency (f)	20	80	190	250	280	170	10

One method of displaying the data in table 1 is shown in Fig. 1, in which heights are indicated horizontally and frequencies are indicated vertically. Consider the first group; the range is 62–64 in. (A to C in the figure), and the frequency is 20. Erect a rectangle $ABDC$ on AC as base and height AB equal to the frequency. Repeat this construction for each of the groups, thus obtaining a

† A. S. Eddington, *Mon. Not. R. Astr. Soc.* **84**, 308, 1924; see also *Internal Constitution of the Stars* (Cambridge University Press, 1926), p. 116.

‡ When we are dealing with observations of, say, a particular characteristic of a star or of the measurement of a physical quantity, the term corresponding to frequency is called the *weight*, usually denoted by w, with the appropriate suffix.

series of rectangles. Since the bases of the rectangles are equal the ratios of the rectangular areas are equivalent to the ratios of the frequencies.

Such a diagram is called a *histogram* and it gives a condensed, although not wholly complete, picture of the distribution of heights amongst the thousand soldiers.

A second method of illustrating the distribution of heights consists in drawing, in Fig. 1, straight lines between the successive mid-points $P_1, P_2, \ldots$ of the upper horizontal sides BD, FG, ... of the rectangles

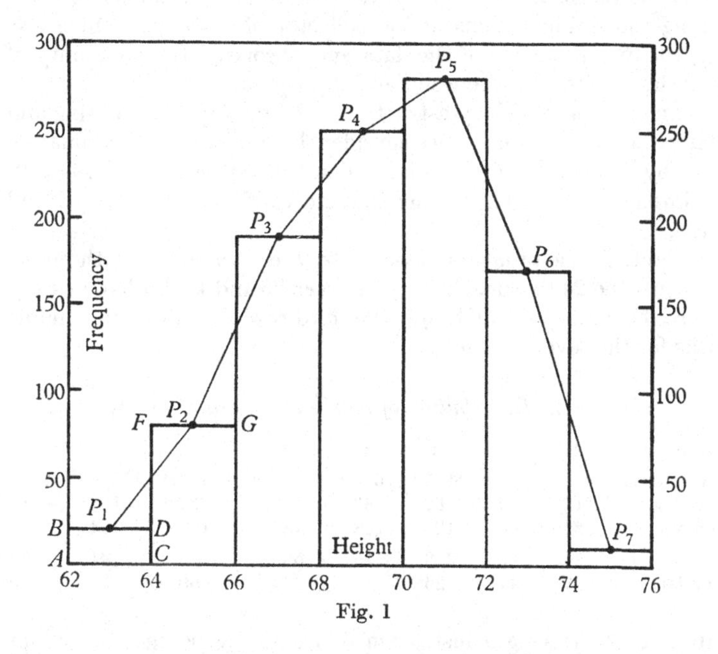

Fig. 1

forming the histogram. The succession of straight lines $P_1P_2, P_2P_3, \ldots$ is called a *frequency polygon*; the coordinates of each of the vertices $P_1, P_2, \ldots$ relate the frequency in a group to the middle height for that group. Like the histogram, the frequency polygon gives a reliable, although not wholly complete, representation of the distribution of heights.

In Fig. 1 the width of each rectangle is 2 in.; this is called the *class interval*, which we denote by c.

For any set of statistics the choice of class interval is dictated in practice according to circumstances, the principal factor being the total frequency—in our example this is the total number of soldiers furnishing the statistics, namely, one thousand. If we had required a more detailed picture of the distribution of heights, we could have taken the class interval to be 1 in., in which event we would derive the frequencies corresponding to height groups 62–63, 63–64, ... The histogram and the frequency polygon would then be constructed according to the principles illustrated in Fig. 1.

1·03. Illustration of the practical application of statistics

The mathematical developments in the theory of statistics have become so recondite in recent years that there is a danger of overlooking the ultimate aims of the collector of statistics, which are, first, the interpretation of the data and, secondly, the possibility of using the results for some well-defined purpose.

As an example which is instructive—and may be illuminating and comforting to the student—we consider the results of an examination taken by 800 candidates and for which the maximum mark is 200; for simplicity, we shall assume that no candidate has scored full marks.

In Table 2† the numbers, denoted by f, of candidates with marks between 0 and 24 (both inclusive), between 25 and 49 (both inclusive), ..., are shown in the fourth row, the third row containing the middle marks for the several groups.

Table 2. *Distribution of marks in an examination*

Group ...	1	2	3	4	5	6	7	8
Range of marks	0–24	25–49	50–74	75–99	100–124	125–149	150–174	175–199
Middle mark	12	37	62	87	112	137	162	187
Frequency (f)	24	64	120	168	184	160	64	16
U	24	88	208	376	560	720	784	800
V (% of U)	3	11	26	47	70	90	98	100

Instead of drawing a histogram or frequency polygon we adopt a third method which consists of constructing a curve, known as an *ogive* curve, based on the numbers, U, of candidates who obtain less than 25, 50, 75, ... marks; these numbers are shown in the fifth row of the table. Thus, the candidates who *fail* to score 50 marks consist of those in groups 1 and 2, and the number, U, in this case is $24+64$ or 88; similarly, the candidates who fail to score 75 marks consist of those in groups 1, 2 and 3, and the number is $88+120$ or 208; and so on.

† To enable the reader to concentrate on principles rather than on arithmetical details the frequencies, f, in an actual examination have been rounded off to the nearest multiple of 8 so that the percentages in the last line of the table are integers.

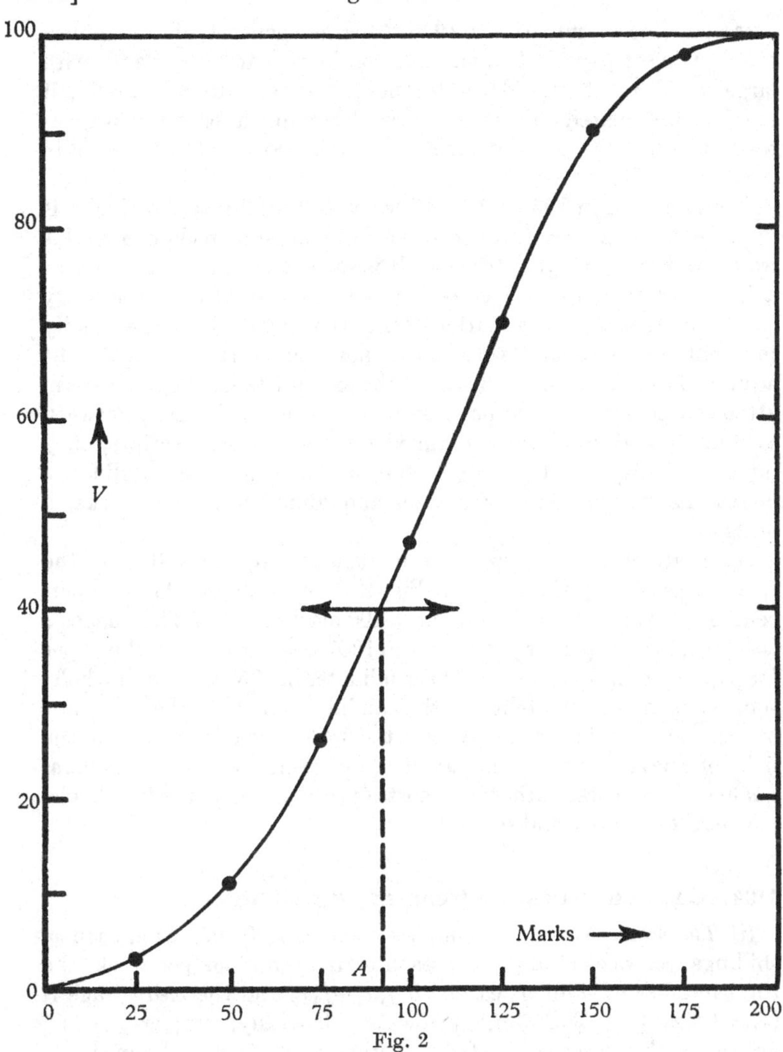

Fig. 2

The last row gives the percentage, V, of candidates failing to score 25, 50, ... marks.

In Fig. 2 marks are indicated on the horizontal axis and the values of V on the vertical axis; the values of V in Table 2 are plotted against the corresponding marks 25, 50, 75, ...

Suppose that the statistics in Table 2, together with Fig. 2, refer to a final degree examination in chemistry for the year 1957; we assume, further, that a similar treatment of marks was made for the corresponding examination in 1956. Now, as a general rule, examiners attempt to set papers of equal difficulty from year to year and, with large numbers of candidates taking the examination annually, it may be anticipated that under normal circumstances† the percentage of failures would remain substantially constant from year to year.

Suppose that, in 1956, 40 % of the candidates failed and that it is desirable to keep the standard of performance—as reckoned in this way—uniform from year to year. It is seen at once from Fig. 2 that a failure of 40 % in 1957 corresponds to the abscissa indicated by the point A, that is, to a mark of 92. If the standard is to be blindly followed, all candidates with marks not greater than 92 would be adjudged to fail. In actual practice the examiners would give careful attention to all 'border-line' candidates with marks, say, between 87 and 99, and, following a detailed scrutiny of their scripts, they might feel disposed to readjust the marks of several candidates, increasing the marks in some cases and diminishing the marks in others.

Alternatively, if the pass mark in 1956 was 100 with 40 % of the candidates failing, the curve in Fig. 2 shows that in 1957 the percentage of candidates failing to reach 100 marks is 47 %. This suggests either that the paper (or papers) set in 1957 was harder than the paper (or papers) set in 1956, or that the candidates in 1957 were, as a whole, somewhat inferior in ability to the candidates in 1956. The business of the examiners in such an event is to effect some compromise in the light of relevant information available to them, such as the general quality of the work, authentic reports of serious epidemics in schools, teachers' estimates, and so on.

1·04. Characteristics of a frequency distribution

(i) *The mean.* We take a simple example. If f_1 men each earn x_1 shillings per week and f_2 men each earn x_2 shillings per week, the total number of shillings earned is $f_1x_1+f_2x_2$ and the *mean* wage is $(f_1x_1+f_2x_2)/(f_1+f_2)$ shillings per week. Generally, if $x_1, x_2, \ldots, x_n$ are the weekly wages of $f_1, f_2, \ldots, f_n$ men respectively, the mean, $\bar{x}$, is given by

$$\bar{x}=\frac{f_1x_1+f_2x_2+\ldots+f_nx_n}{f_1+f_2+\ldots+f_n}=\frac{\sum_{i=1}^{n} f_ix_i}{\sum_{i=1}^{n} f_i}. \tag{1}$$

† *Abnormal* circumstances would include, for example, the partial dislocation of education, to different degrees, in different parts of the country due to large-scale illness or, as was found between 1939 and 1945, to war-time conditions.

Here we regard x_1, x_2, ... as the particular values of a *variable* x, usually referred to in the theory of statistics as the *variate*.

Let N denote the total number of men concerned—or, in statistical language, the total frequency. Then

$$N = f_1 + f_2 + \dots + f_n = \sum_{i=1}^{n} f_i$$

and (1) becomes
$$\bar{x} = \frac{1}{N} \sum_{1}^{n} f_i x_i. \tag{2}$$

This formula is applicable in all such problems where the frequencies associated with particular values of the variable are given.

(ii) *The median.* The median is the value, m, of the variable dividing the distribution of statistics such that the frequency of values less than m is equal to the frequency of values greater than m.

Referring to Table 2 (p. 4) we see from the values of U that 376 candidates have marks less than 100—that is, up to 99—and that 560 candidates have marks less than 125—that is, up to 124. It is evident that the median, m, lies between 99 and 124 marks. Moreover, we could ascertain the median mark by examining the individual marks of the 184 candidates in group 5 (marks between 100 and 124) by arranging these in increasing numerical sequence and noting the mark not exceeded by the first 24 candidates in the group, 24 being the balance between 376 and 400.

This procedure is tedious and in practice we proceed as follows. If it is assumed that the marks in group 5 are uniformly distributed in the range 100–124, we have to find the mark, m, such that 24 out of the 184 candidates in the group have this mark. Then

$$m = 99 + \tfrac{24}{184} . 25 = 102{\cdot}3,$$

or, to the nearest integer, $m = 102$.

This result is verifiable from Fig. 2, being the mark associated with 50 % of the candidates.

(iii) *The quartiles.* The quartiles are the three values of the variable—q_1, q_2 and q_3—such that the frequencies for values of the variable between 0 and q_1, between q_1 and q_2, between q_2 and q_3, and between q_3 and N (the total frequency over the whole range) are all equal, each being $\frac{1}{4}N$. It is evident that q_2 is the median.

The practical method of calculating q_1 and q_3 is the same as that for calculating the median.

In the theory of errors the quartiles q_1 and q_3 have a special significance (see § 4·09).

(iv) *The mode.* The mode is the value of the variable for which the frequency is a maximum.

Table 1 shows that the maximum frequency in the various groups is 280—in group 5—corresponding to the middle height 71; this last number is the abscissa of the point, P_5, in the frequency polygon (Fig. 1) corresponding to the frequency 280; accordingly, the mode is 71.

If we had taken the class interval to be, say, a half of that to which Table 1 refers, the value of the mode found in this way may be expected to differ slightly from the first value 71; the value of the mode is thus dependent to some extent on the selection of the class interval.

1·05. Calculation of the mean

By 1·04 (2), the mean $\bar{x}$ of the n discrete values $x_1, x_2, \ldots, x_n$ of a variable x, with frequencies $f_1, f_2, \ldots, f_n$, is given by

$$\bar{x} = \frac{1}{N} \sum_{i=1}^{n} f_i x_i. \tag{1}$$

Considering our example in § 1·03 we can find $\bar{x}$ *accurately* by forming the sum in (1) by means of the frequencies associated with the individual marks 0, 1, 2, ..., 199. This would be an intolerably long and tedious calculation.

We can condense the calculation and obtain a sufficiently reliable value of $\bar{x}$ by assuming that, in the first group (Table 2, p. 4), the average mark of the 24 candidates in the group is the middle mark 12 for the range 0–24 marks; then $x_1 = 12$ and $f_1 = 24$. Similarly, we assume that the average mark of the 64 candidates in group 2 is the middle mark 37; then, $x_2 = 37$ and $f_2 = 64$. The remaining groups are treated in a similar way. Our calculation for $\bar{x}$, by means of (1), would then be

$$\bar{x} = \frac{1}{N}[24.12 + 64.37 + \ldots + 16.187],$$

where $N = 800$. It can be verified that $\bar{x} = 100\frac{3}{4}$, or, to the nearest integer, $\bar{x} = 101$. The calculation is still long and tedious.

To simplify the arithmetical work still further we introduce the following device. Let a denote a convenient value of the variable which we estimate to be in the neighbourhood of $\bar{x}$. Let

$$x_i = a + \xi_i, \tag{2}$$

from which the values of ξ_i (which can be positive or negative) are readily derived. Then,† by (1) and (2),

$$N\bar{x} = \Sigma f_i(a + \xi_i) = a\Sigma f_i + \Sigma f_i \xi_i. \tag{3}$$

† In (3) and elsewhere the limits 1 to n in the summations will be omitted for simplicity when no confusion is likely to be caused.

Let $\bar{\xi}$ denote the mean of the quantities ξ_i with frequencies f_i; then $N\bar{\xi}=\Sigma f_i\xi_i$, and (3) becomes

$$N\bar{x}=Na+N\bar{\xi}$$

or

$$\bar{x}=a+\bar{\xi}. \tag{4}$$

This formula is the appropriate one when the class interval is unity. When the class interval, c, is different from unity, we can simplify the computations still further by writing

$$\xi_i=cu_i, \tag{5}$$

from which $\Sigma f_i\xi_i=c\Sigma f_iu_i$, so that

$$\bar{\xi}=c\bar{u}, \tag{6}$$

where $\bar{u}$ is the mean of the quantities u_i with associated frequencies f_i. Formula (4) then becomes

$$\bar{x}=a+c\bar{u}. \tag{7}$$

We use the statistics of Table 2 to calculate $\bar{x}$ first by means of (4) and, secondly, by means of (7); the details are found in Table 3, the second column of which gives the middle mark of each of the groups and the third column gives the corresponding frequencies.

Table 3. *Calculation of the mean*

(1) Group	(2) x_i (middle mark)	(3) f_i	(4) ξ_i	(5) $f_i\xi_i$	(6) u_i	(7) f_iu_i
1	12	24	−100	− 2,400	−4	− 96
2	37	64	− 75	− 4,800	−3	−192
3	62	120	− 50	− 6,000	−2	−240
4	87	168	− 25	− 4,200	−1	−168
5	112	184	0	**−17,400**	0	**−696**
6	137	160	+ 25	+ 4,000	+1	+160
7	162	64	+ 50	+ 3,200	+2	+128
8	187	16	+ 75	+ 1,200	+3	+ 48
		800		**+ 8,400**		**+336**

Summary: $N=800$; $\Sigma f\xi=-9000$; $\Sigma fu=-360$.

As a rough guess the mean mark $\bar{x}$ is between 87 and 112 (groups 4 and 5) and, since from column 2 the class interval, c, is 25, it is evident from (5) that it would be convenient to have the values of ξ_i given by multiples of 25, for then the values of u_i will be integers. A value of a satisfying these desiderata is clearly 112.

In column 4 are to be found the values of ξ_i and in the next column the values of $f_i \xi_i$; the sum of the negative values of $f_i \xi_i$ is shown near the middle of column 5 and the sum of the positive values at the bottom of the column; these values are $-17{,}400$ and $+8{,}400$ respectively; the final sum is then -9000, as shown in the summary at the foot of the table. Hence, by (4), since the total frequency, N, is 800,

$$\bar{x} = 112 + \tfrac{1}{800}(-9000) = 100\tfrac{3}{4},$$

or, to the nearest integer, $\bar{x} = 101$, agreeing with the result previously stated.

In column 6 the values of u_i, based on (5), are given and in the last column the values of $f_i u_i$ are found, with the negative and positive sums shown as in column 5. The value of $\bar{u}$ is given by

$$\bar{u} = \tfrac{1}{800}(-696 + 336) = -\tfrac{9}{20},$$

and hence, by (7),

$$\bar{x} = 112 - 25 . \tfrac{9}{20} = 100\tfrac{3}{4},$$

or, to the nearest integer, $\bar{x} = 101$.

The advantages of the second method (that involving $\bar{u}$) over the first method (involving $\bar{\xi}$) as regards simplicity and economy of calculation are sufficiently obvious to require no further emphasis.

1·06. Moments

Consider the frequency polygon in Fig. 3 with n vertices $P_1, P_2, \ldots, P_n$, and, in particular, the ith vertex, P_i, corresponding to the value, x_i, of the variable; in the figure, $OQ_i = x_i$ and $Q_i P_i = f_i$. Let AB be any line parallel to OY and denote the abscissa of A by a.

The r-th moment of the frequency distribution about AB—that is, about the line $x = a$—is denoted by $\mu_r(a)$ and defined by

$$\mu_r(a) = \frac{1}{N} \Sigma f_i (x_i - a)^r, \tag{1}$$

r being a positive integer, including zero.

Write, as before,

$$\xi_i = x_i - a. \tag{2}$$

Then, as in 1·05 (4)

$$\bar{\xi} = \bar{x} - a. \tag{3}$$

From (1) we then have, by means of (2),

$$\mu_r(a) = \frac{1}{N} \Sigma f_i \xi_i^r. \tag{4}$$

We refer specifically to the algebraical quantity $\xi_i \equiv x_i - a$ as the *deviation* of x_i from a; the values of ξ_i may be positive or negative.

The principal moments are those taken about the mean of the n values x_i, that is, about the line $x=\bar{x}$. The principal moments are denoted simply by μ_r, so that

$$\mu_r=\frac{1}{N}\Sigma f_i(x_i-\bar{x})^r. \tag{5}$$

The line $x=\bar{x}$ is usually referred to as the *centroid vertical*.

In particular, from (5) and (4) when $r=0$, we have

$$\mu_0=\mu_0(a)=\frac{1}{N}\Sigma f_i;$$

hence

$$\mu_0=\mu_0(a)=1. \tag{6}$$

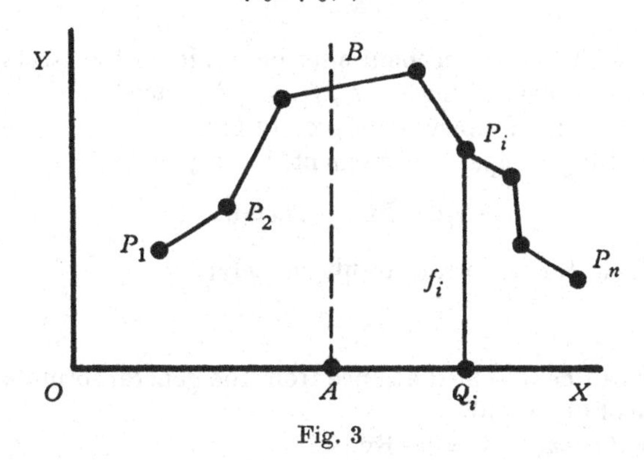

Fig. 3

In general, the formula (5) becomes, by means of (2) and (3),

$$\mu_r=\frac{1}{N}\Sigma f_i(\xi_i-\bar{\xi})^r, \tag{7}$$

or, on expansion,

$$\mu_r=\frac{1}{N}\left[\Sigma f_i\xi_i^r-\frac{r}{1!}\bar{\xi}\Sigma f_i\xi_i^{r-1}+\frac{r(r-1)}{2!}\bar{\xi}^2\Sigma f_i\xi_i^{r-2}-\dots\right].$$

Hence, by (4),

$$\mu_r=\mu_r(a)-\frac{r}{1!}\bar{\xi}\mu_{r-1}(a)+\frac{r(r-1)}{2!}\bar{\xi}^2\mu_{r-2}(a)-\dots, \tag{8}$$

the general term being

$$(-1)^k\frac{r(r-1)\dots(r-k+1)}{k!}\bar{\xi}^k\mu_{r-k}(a).$$

The calculation of μ_r is most easily achieved by first calculating the moments about $x=a$, where a is conveniently chosen, and then applying (8).

(i) *First moments* ($r=1$). By (1),

$$\mu_1(a)=\frac{1}{N}\Sigma f_i(x_i-a)=\frac{1}{N}\Sigma f_i x_i-\frac{a}{N}\Sigma f_i;$$

but $\Sigma f_i x_i = N\bar{x}$ and $\Sigma f_i = N$; hence

$$\mu_1(a)=\bar{x}-a=\bar{\xi}. \tag{9}$$

In particular, if $a=0$, then $\mu_1(0)=\bar{x}$, that is, the first moment about the line $x=0$ is the mean, $\bar{x}$. The formula

$$\mu_1(0)\equiv\frac{1}{N}\Sigma f_i x_i=\bar{x}$$

is identical with that in mechanics for determining the x-coordinate of the centre of mass of masses $f_1, f_2, \ldots, f_n$ placed at the vertices $P_1, P_2, \ldots, P_n$ of the frequency polygon in Fig. 3.

From (5), the principal first moment, μ_1, is given by

$$N\mu_1=\Sigma f_i x_i-N\bar{x}=0.$$

Hence we have the important result, namely,

$$\mu_1=0. \tag{10}$$

This result can be obtained at once from the general formula (8) on making use of (9) and (6).

(ii) *Second moments* ($r=2$). By (1),

$$N\mu_2(a)=\Sigma f_i(x_i-a)^2=\Sigma f_i\xi_i^2. \tag{11}$$

This formula is identical with that in mechanics for determining the moment of inertia about AB of masses $f_1, f_2, \ldots, f_n$ placed at $P_1, P_2, \ldots, P_n$ in Fig. 3.

The formula (11) shows that $\mu_2(a)$ is the mean of the squares of the deviations of the variable with respect to a; $\mu_2(a)$ is referred to as the *mean square deviation*.

From (8),

$$\mu_2=\mu_2(a)-2\bar{\xi}\mu_1(a)+\bar{\xi}^2\mu_0(a),$$

or, by (9) and (6),

$$\mu_2=\mu_2(a)-\bar{\xi}^2 \tag{12}$$

or

$$\mu_2=\mu_2(a)-(\bar{x}-a)^2. \tag{13}$$

(iii) *Variance and standard deviation.* The principal second moment, μ_2, is called the *variance* and is denoted by the special symbol, σ^2; thus, from (13),

$$\sigma^2\equiv\mu_2=\mu_2(a)-(\bar{x}-a)^2. \tag{14}$$

Also, σ is called the *standard deviation*.

The variance and the standard deviation are important quantities in the theory of statistics.

(iv) *Third moments* ($r=3$). By (1) and (4),

$$N\mu_3(a)=\Sigma f_i(x_i-a)^3=\Sigma f_i\xi_i^3.$$

Again, by (8), $\mu_3=\mu_3(a)-3\bar{\xi}\mu_2(a)+3\bar{\xi}^2\mu_1(a)-\bar{\xi}^3,$

or, by (9),
$$\mu_3=\mu_3(a)-3\bar{\xi}\mu_2(a)+2\bar{\xi}^3. \tag{15}$$

(v) *Fourth moments* ($r=4$). By (1) and (4),

$$N\mu_4(a)=\Sigma f_i(x_i-a)^4=\Sigma f_i\xi_i^4.$$

Again, by (8),

$$\mu_4=\mu_4(a)-4\bar{\xi}\mu_3(a)+6\bar{\xi}^2\mu_2(a)-4\bar{\xi}^3\mu_1(a)+\bar{\xi}^4,$$

or, by (9),
$$\mu_4=\mu_4(a)-4\bar{\xi}\mu_3(a)+6\bar{\xi}^2\mu_2(a)-3\bar{\xi}^4. \tag{16}$$

The formula for the principal moments of higher order, if required, can be derived in a similar way.

1·07. Formulae for the moments in terms of the class interval c

Write
$$\xi_i\equiv x_i-a=cu_i.$$

Then
$$N\mu_r(a)\equiv\Sigma f_i(x_i-a)^r=c^r\Sigma f_iu_i^r.$$

It is convenient to denote the rth moment of the quantities u_i with respect to a by $\mu'_r(a)$ so that $\mu'_r(a)=\Sigma f_iu_i^r$. Hence

$$\mu_r(a)=c^r\mu'_r(a). \tag{1}$$

The procedure is then to derive the numerical values of the moments $\mu'_r(a)$, from which the values of the moments $\mu_r(a)$ are at once found by means of (1).

The principal moments can then be obtained by means of the formulae in the previous section.

1·08. Example of the calculation of moments

We take as an example the following statistics relating to the wages of 300 workmen in a factory: 8 men each earn 125*s*. weekly, 24 men each earn 135*s*. weekly and so on. The value, x, of the weekly wage in each of 8 groups is given in the first column of Table 4, with the frequency f in the second column.

From column 1, the class interval is 10; it is then convenient to take a as 165; the deviations, ξ, from a are given in column 3; the values of u ($\equiv\xi/c$) are given in column 4 and these are positive or negative integers. The entries in columns 5–8 are then inserted. Columns 9 and 10 will be discussed later.

Table 4. *Distribution of wages*

(1)	(2)	(3)	(4)	(5)	(6)	(7)	(8)	(9)	(10)
x	f	ξ	u	fu	fu^2	fu^3	fu^4	η	$f\eta$
125	8	−40	−4	− 32	128	− 512	2048	36	288
135	24	−30	−3	− 72	216	− 648	1944	26	624
145	52	−20	−2	−104	208	− 416	832	16	832
155	70	−10	−1	− 70	70	− 70	70	6	420
165	58	0	0	**−278**	0	**−1646**	0	4	232
175	39	+10	+1	+ 39	39	+ 39	39	14	546
185	28	+20	+2	+ 56	112	+ 224	448	24	672
195	21	+30	+3	+ 63	189	+ 567	1701	34	714
	300			**+158**	**962**	**+ 830**	**7082**		**4328**

Summary: $N=300$; $\Sigma fu=-278+158=-120$; $\Sigma fu^2=962$;

$\Sigma fu^3=-1646+830=-816$; $\Sigma fu^4=7082$; $\Sigma f\eta=4328$.

The principal information arising from the statistics is summarized at the foot of the table.

(i) *The mean*

$$\bar{u}=\frac{1}{N}\Sigma f_i u_i=-\tfrac{120}{300}=-\tfrac{2}{5};$$

hence, since $c=10$ and $a=165$,

$$\bar{\xi}=-4 \quad\text{and}\quad \bar{x}\equiv a+\bar{\xi}=161.$$

(ii) *The variance* $\quad \mu_2'(a)\equiv\dfrac{1}{N}\Sigma f_i u_i^2=\tfrac{962}{300};$

hence $\quad \mu_2(a)\equiv c^2\mu_2'(a)=320\tfrac{2}{3}.$

By 1·06 (14), $\quad \mu_2\equiv\sigma^2=\mu_2(a)-\bar{\xi}^2=304\tfrac{2}{3}.$

The standard deviation is then given by $\sigma=17{\cdot}46$.

(iii) *Principal third moment*

$$\mu_3'(a)\equiv\frac{1}{N}\Sigma f_i u_i^3=-\tfrac{816}{300};$$

hence $\quad \mu_3(a)\equiv c^3\mu_3'(a)=-2720.$

By 1·06 (15),

$$\mu_3=-2720-3(-4).320\tfrac{2}{3}+2(-4)^3=1000.$$

(iv) *Principal fourth moment*

$$\mu_4'(a)\equiv\frac{1}{N}\Sigma f_i u_i^4=\tfrac{7082}{300};$$

hence $\quad \mu_4(a)\equiv c^4\mu_4'(a)=236066\tfrac{2}{3}.$

By 1·06 (16) it is then found that

$$u_4 = 222562\tfrac{2}{3}.$$

(v) *Average deviation from the mean.* Let η_i denote the numerical value—that is, independent of sign—of the deviation of x_i from the mean so that $\eta_i = |x_i - \bar{x}|$; we refer to η_i as the *numerical deviation.* The values of η_i and $f_i\eta_i$ are given in columns 9 and 10.

Let η denote the mean of the values η_i; then

$$\eta = \frac{1}{N}\Sigma f_i\eta_i = \tfrac{4328}{300} = 14{\cdot}43.$$

We refer to η as the *average deviation.*

(vi) *Summary of results.* The values of the mean, the principal moments and the average deviation are all characteristics of the given distribution. The values are collected here for future reference:

$$\bar{x} = 161; \quad \mu_2 = 304\tfrac{2}{3}; \quad \mu_3 = 1000;$$

$$\mu_4 = 222562\tfrac{2}{3}; \quad \eta = 14{\cdot}43.$$

1·09. Skewness and kurtosis

(i) The *skewness* of a frequency distribution is a particular measure of its departure from symmetry. In a symmetrical distribution, the mean and the mode are identical; also, $\mu_1 = 0$ and $\mu_3 = 0$, the latter following from symmetry. In a non-symmetrical distribution we have $\mu_1 = 0$, by 1·06 (10), but μ_3 is no longer zero and can have a positive or negative value.

One measure of skewness encountered in statistical theory is defined in terms of the quantity μ_3/σ^3 which is denoted by $\beta_1^{\frac{1}{2}}$, so that

$$\beta_1^{\frac{1}{2}} = \mu_3/\sigma^3. \tag{1}$$

Also

$$\beta_1 \equiv \frac{\mu_3^2}{\sigma^6} = \frac{\mu_3^2}{\mu_2^3} = \frac{(\mu_3')^2}{(\mu_2')^3}; \tag{2}$$

thus β_1 is independent of the class interval. The skewness, $\beta_1^{\frac{1}{2}}$, is positive or negative according as μ_3 is positive or negative.

For the distribution in Table 4, as summarized in 1·08 (vi), we found: $\mu_2 \equiv \sigma^2 = 304\tfrac{2}{3}$ and $\mu_3 = 1000$; hence

$$\beta_1^{\frac{1}{2}} = +0{\cdot}594. \tag{3}$$

Pearson's definition of skewness, which is in frequent use, is as follows:

$$\text{Skewness} = \frac{\text{Mean} - \text{Mode}}{\sigma}. \tag{4}$$

From Table 4 the mode is 155 (the value of x corresponding to the maximum frequency); also, $\bar{x}=161$ and $\sigma=17{\cdot}46$. Hence

$$\text{Skewness}=\frac{161-155}{17{\cdot}46}=+0{\cdot}344.$$

The two numerical results for the skewness, found above, are not necessarily comparable, since the two definitions of skewness are expressed in terms of different functions of the principal quantities associated with the frequency distribution.

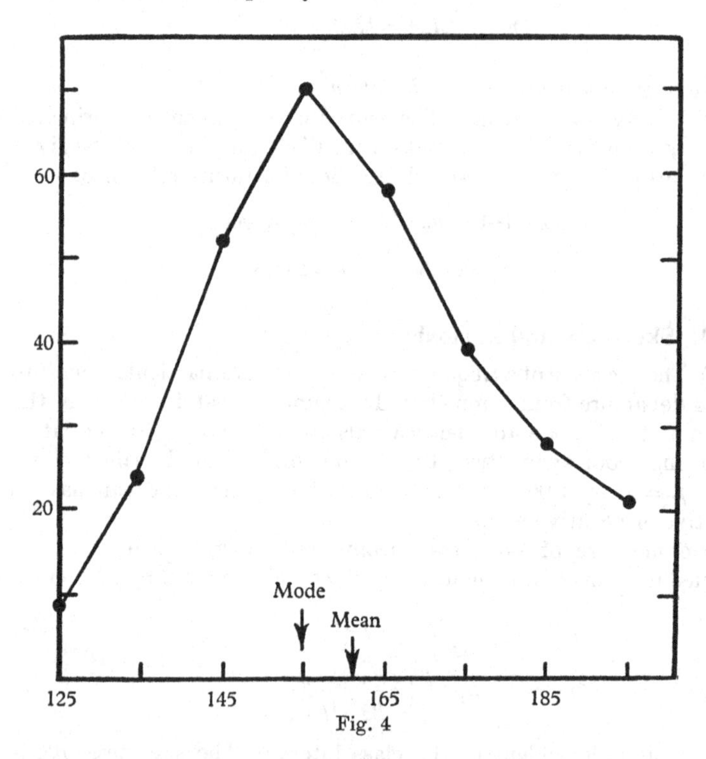

Fig. 4

In definition (4) the skewness is said to be positive or negative according as the mean is greater or less than the mode. The sign of the skewness thus gives an indication as to one characteristic of the frequency distribution, namely, the relative positions of the mean and the mode; it is, in consequence, more illuminating than the definition according to the value of $\beta_1^{\frac{1}{2}}$ in (1).

The frequency polygon relating to the statistics in Table 4 is shown in Fig. 4, with the mean and the mode indicated; the skewness in this case is positive according to Pearson's definition.

(ii) *Kurtosis*. This is a measure of the characteristic of a frequency distribution which involves the fourth principal moment; it is defined by β_2, given by

$$\beta_2 = \frac{\mu_4}{\mu_2^2} = \frac{\mu_4'}{(\mu_2')^2}; \tag{5}$$

β_2 is independent, like β_1, of the class interval.

It is easily found from the values of μ_2 and μ_4 in 1·08 (vi) that $\beta_2 = 2{\cdot}40$.

As we shall see later, in § 115 (v), the value of β_2 for a *normal distribution* is 3; in our example the distribution is said to have a *defect of kurtosis*, measured by $\beta_2 - 3$ or $-0{\cdot}60$.

1·10. Relative frequency

We assume, as in previous sections, that the statistics are condensed to give the frequency in a class interval c so that f_i is the frequency corresponding to values of the variable between $x_i - \frac{1}{2}c$ and $x_i + \frac{1}{2}c$; the corresponding rectangle in the histogram has a base c and height f_i. If N is the total frequency, then f_i/N is the *relative frequency*.

1·11. Continuous frequency distributions

We consider the statistics of a variable x between the values $x = \alpha$ and $x = \beta$ $(\beta > \alpha)$, the total frequency being N. As we have seen, the class interval c is selected in practice with due regard to the magnitude of N; for example, in Table 2, $c = 25$ and $N = 800$; if the total frequency had been, say 10,000, we could with advantage have taken a much smaller value of c, thereby gaining from the histogram or the frequency polygon a more accurate view of the characteristics of the distribution. In such a case, unless the distribution is peculiar, the vertices of the frequency polygon may be expected to suggest a continuous curve, the character of which becomes more definite as N is still further increased and c diminished. When N is very large and c is very small, it will be assumed that the vertices of the frequency polygon lie on a continuous curve whose equation can, in principle, be determined (see Chapter 7); we take the equation of the curve to be

$$y = F(x), \tag{1}$$

which defines a *continuous frequency distribution*.

In Fig. 5, AB is the curve represented by (1) for values of x between α and β, OA_1 being α and OB_1 being β; P is any point, with coordinates (x, y), on AB and QP is the corresponding ordinate; also, $OQ = x$.

CD is an infinitesimal class interval, δx, with Q as its mid-point so that $OC = x - \frac{1}{2}\delta x$ and $OD = x + \frac{1}{2}\delta x$.

If we assume that P is a vertex of the frequency polygon, the frequency for the class interval CD is $QP.\delta x$ or $y\,\delta x$ or $F(x)\,\delta x$. The total frequency, which we denote by S, is given by $S = \Sigma F(x)\,\delta x$ for the range $\alpha \leqslant x \leqslant \beta$, or

$$S = \int_{\alpha}^{\beta} F(x)\,dx; \qquad (2)$$

thus, S is the area under the curve AB.

It is usual to replace the infinitesimal δx by the differential dx, so that we write:

the frequency for values of x between $x - \tfrac{1}{2}dx$ and $x + \tfrac{1}{2}dx$ is $F(x)\,dx$,

or, in the form which we shall generally use,

the frequency for values of x between x and $x + dx$ is $F(x)\,dx$.

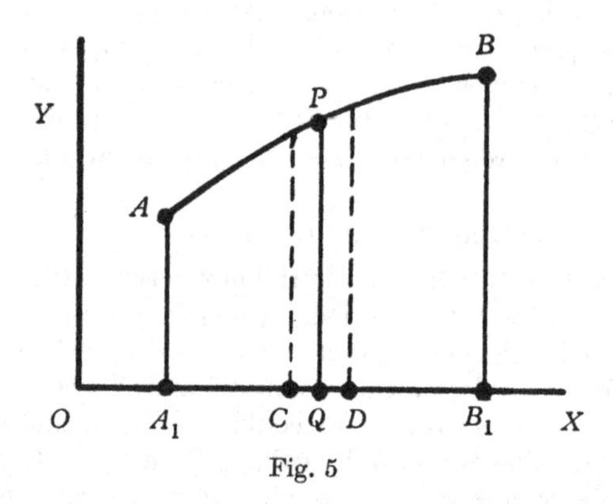

Fig. 5

The *relative frequency* for values of x between x and $x + dx$ is then $\dfrac{1}{S} F(x)\,dx$.

The relative frequency for values of the variable between γ and δ $(\alpha < \gamma < \delta < \beta)$ is $\dfrac{1}{S}\displaystyle\int_{\gamma}^{\delta} F(x)\,dx$.

Ex. If $y = 1 + x^2$, the range of values of x being from 2 to 5, then

$$S = \int_{2}^{5} (1 + x^2)\,dx = [x + \tfrac{1}{3}x^3]_2^5 = 42.$$

The relative frequency for values of the variable between x and $x + dx$ is then $\frac{1}{42}(1 + x^2)\,dx$.

The relative frequency for values of the variable between 3 and 4 is $\dfrac{1}{42}\displaystyle\int_{3}^{4} (1 + x^2)\,dx = \tfrac{1}{42}[x + \tfrac{1}{3}x^3]_3^4 = \tfrac{20}{63}$.

1·12. Moments for continuous frequency distributions

(i) In terms of discrete values of a variable x, the r-th moment about the line $x=a$ is, as in 1·06 (i), defined by

$$\mu_r(a)=\frac{1}{N}\Sigma f_i(x_i-a)^r,$$

where N is the total frequency.

For a continuous distribution given by $y=F(x)$ between the values $x=\alpha$ and $x=\beta$, the r-th moment of the frequency $F(x)\,dx$ for values of the variable between x and $x+dx$ about the line $x=a$ is

$$F(x)\,dx\,(x-a)^r.$$

Hence
$$\mu_r(a)=\frac{1}{S}\int_\alpha^\beta (x-a)^r F(x)\,dx, \tag{1}$$

where S is the total frequency given by

$$S=\int_\alpha^\beta F(x)\,dx. \tag{2}$$

(ii) *Mean value*, $\bar{x}$. The mean value, $\bar{x}$, is the first moment about the origin; hence, from (1), on putting $r=1$ and $a=0$,

$$\bar{x}\equiv\mu_1(0)=\frac{1}{S}\int_\alpha^\beta xF(x)\,dx. \tag{3}$$

Frequently, the form of $F(x)$ suggests that it would be advantageous to develop a formula for $\bar{x}$ in terms of $\mu_1(a)$, where a has a particular relevance. From (1)

$$\mu_1(a)=\frac{1}{S}\int_\alpha^\beta (x-a)\,F(x)\,dx=\frac{1}{S}\int_\alpha^\beta xF(x)\,dx-\frac{a}{S}\int_\alpha^\beta F(x)\,dx$$

$$=\bar{x}-a, \quad \text{from (3) and (2);}$$

hence
$$\bar{x}=a+\mu_1(a). \tag{4}$$

Ex. $F(x)=(x-a)^4;\quad \alpha=a,\ \beta=2a.$

Then

$$S=\int_a^{2a}(x-a)^4\,dx=[\tfrac{1}{5}(x-a)^5]_a^{2a}=\tfrac{1}{5}a^5.$$

$$\mu_1(a)\equiv\frac{1}{S}\int_a^{2a}(x-a)\,F(x)\,dx=[\tfrac{5}{6}a^5(x-a)^6]_a^{2a}=\tfrac{5}{6}a.$$

Hence, from (4), $\bar{x}\equiv a+\mu_1(a)=\tfrac{11}{6}a.$

(iii) *The variance and standard deviation.* We have, for the principal moment of the second order,

$$\sigma^2 \equiv \mu_2 = \frac{1}{S}\int_\alpha^\beta (x-\bar{x})^2 F(x)\,dx$$

$$= \frac{1}{S}\int_\alpha^\beta x^2 F(x)\,dx - 2\frac{\bar{x}}{S}\int_\alpha^\beta xF(x)\,dx + \frac{\bar{x}^2}{S}\int_\alpha^\beta F(x)\,dx;$$

hence, by means of (3) and (2),

$$\sigma^2 = \mu_2(0) - \bar{x}^2. \tag{5}$$

This formula enables us to calculate the variance and then the standard deviation.

As in (ii) it may be more convenient to derive σ^2 in terms of moments about the line $x=a$. Then

$$S\mu_2(a) = \int_\alpha^\beta (x-a)^2 F(x)\,dx$$

$$= \int_\alpha^\beta x^2 F(x)\,dx - 2a\int_\alpha^\beta xF(x)\,dx + a^2\int_\alpha^\beta F(x)\,dx$$

$$= S\mu_2(0) - 2aS\bar{x} + a^2 S;$$

hence, by (5), $$\mu_2(a) = \sigma^2 + \bar{x}^2 - 2a\bar{x} + a^2,$$

from which $$\sigma^2 = \mu_2(a) - (\bar{x}-a)^2. \tag{6}$$

It will be observed that formulae (4) and (6) are the same as the formula 1·06 (9) and 1·06 (13) derived for discrete values of the variable.

Ex. $$F(x) = (x-a)^4; \quad \alpha = a,\ \beta = 2a.$$

$$S\mu_2(a) = \int_a^{2a} (x-a)^6\,dx = \tfrac{1}{7}a^7;$$

hence $$\mu_2(a) = \tfrac{5}{7}a^2,$$

and $$\sigma^2 = \tfrac{5}{7}a^2 - (\bar{x}-a)^2 = \tfrac{5}{7}a^2 - (\tfrac{11}{6}a - a)^2$$

or $$\sigma^2 = \tfrac{5}{252}a^2.$$

(iv) *Principal higher moments.* These are found from the general formulae (1) and (2). For moments about the mean we have

$$\mu_r = \frac{1}{S}\int_\alpha^\beta (x-\bar{x})^r F(x)\,dx.$$

If it is more convenient to calculate the moments about $x=a$, it is easily seen by the procedure adopted in (ii) and (iii) that μ_3 is given by 1·06 (15) and μ_4 by 1·06 (16).

(v) *The average deviation from the mean.* We denote the average deviation from the mean, *taken without regard to sign*, by η; then η is given by

$$\eta=\frac{1}{S}\int_\alpha^\beta |x-\bar{x}|\, F(x)\, dx. \tag{7}$$

Now, in general,

$$\int_\alpha^\beta \phi(x)\, dx=\int_\alpha^{\bar{x}} \phi(x)\, dx+\int_{\bar{x}}^\beta \phi(x)\, dx;$$

hence (7) becomes, since $\bar{x}-\alpha$ is positive in the range $\alpha<x<\bar{x}$ and $\beta-\bar{x}$ is positive in the range $\bar{x}<x<\beta$,

$$\eta=\frac{1}{S}\int_\alpha^{\bar{x}} (\bar{x}-x)\, F(x)\, dx+\frac{1}{S}\int_{\bar{x}}^\beta (x-\bar{x})\, F(x)\, dx.$$

Ex. $F(x)=x^2;\quad \alpha=0,\ \beta=a.$

It is readily found that $S=\frac{1}{3}a^3$ and $\bar{x}=\frac{3}{4}a$.

Then

$$\begin{aligned} S\eta &= \int_0^{\bar{x}} (\bar{x}-x)\, x^2 dx+\int_{\bar{x}}^a (x-\bar{x})\, x^2 dx \\ &= [\tfrac{1}{3}\bar{x}x^3-\tfrac{1}{4}x^4]_0^{\bar{x}}+[\tfrac{1}{4}x^4-\tfrac{1}{3}\bar{x}x^3]_{\bar{x}}^a \\ &= \tfrac{1}{6}\bar{x}^4+\tfrac{1}{4}a^4-\tfrac{1}{3}a^3\bar{x}=\tfrac{1}{6}\bar{x}^4. \end{aligned}$$

Hence $$\eta=\tfrac{81}{512}a.$$

1·13. The normal frequency distribution

An important frequency distribution of the deviations, ξ, of a variable x from the mean, $\bar{x}$, is represented by the *normal function* which we denote in the first instance by $F(\xi)$, expressed by the formula

$$F(\xi)=A\, e^{-h^2\xi^2}\equiv A\, e^{-h^2(x-\bar{x})^2}, \tag{1}$$

in which A and h are constants and $\bar{x}$ is supposed known.

As will be seen in later chapters the normal function is associated very intimately with the theory of errors.

The frequency of the values of the deviations between ξ and $\xi+d\xi$—or the frequency of the values of the variable between x and $x+dx$—is given by

$$A\, e^{-h^2\xi^2} d\xi \quad \text{or} \quad A\, e^{-h^2(x-\bar{x})^2} dx.$$

The deviations, ξ, are assumed to take all values between $-\infty$ and $+\infty$; the total frequency, S, is then given by

$$S=A\int_{-\infty}^{\infty} e^{-h^2x^2}\, dx. \tag{2}$$

It will be shown in §1·14 that the value of the integral in (2) is $\sqrt{\pi}/h$; hence

$$S = A\sqrt{\pi}/h. \tag{3}$$

The *relative frequency* of the deviations between ξ and $\xi + d\xi$ is denoted by $f(\xi)\,d\xi$ and, accordingly,

$$f(\xi)\,d\xi = \frac{1}{S}F(\xi)\,d\xi;$$

hence, from (1) and (3),
$$f(\xi) = \frac{h}{\sqrt{\pi}}e^{-h^2\xi^2}. \tag{4}$$

The function $f(\xi)$, given by (4), is the normal frequency function, and it is important to remember that it is concerned with *relative frequency*.

It is convenient, however, to regard the function $f(\xi)$ to be associated with a very large number, N, of deviations; we can then express the fundamental principle in the form:

$$\text{the frequency of deviations between } \xi \text{ and } \xi + d\xi = N\frac{h}{\sqrt{\pi}}e^{-h^2\xi^2}\,d\xi. \tag{5}$$

1·14. Integrals associated with the normal function

(i) *Proof of* $\int_0^\infty e^{-t^2}\,dt = \tfrac{1}{2}\sqrt{\pi}$.

Let (x, y) be the coordinates of a point referred to axes X_1OX, Y_1OY, Also, let I denote the integral; then, since a definite integral is independent of the variable, we have

$$I = \int_0^\infty e^{-x^2}\,dx = \int_0^\infty e^{-y^2}\,dy$$

and
$$I^2 = \int_0^\infty\int_0^\infty e^{-(x^2+y^2)}\,dx\,dy. \tag{1}$$

Thus, I^2 denotes the double integral taken over the area of the first quadrant XOY, that is, for $0 \leqslant x \leqslant \infty$ and $0 \leqslant y \leqslant \infty$.

Let $x = r\cos\theta$ and $y = r\sin\theta$, the range for r being 0 to ∞ and the range for θ being 0 to $\tfrac{1}{2}\pi$; then the infinitesimal area in polar coordinates is $r\,dr\,d\theta$ and (1) becomes

$$I^2 = \int_0^\infty\int_0^{\frac{1}{2}\pi} e^{-r^2}r\,dr\,d\theta,$$

or, since r and θ are independent,

$$I^2 = \tfrac{1}{4}\pi\int_0^\infty e^{-r^2}\,d(r^2) = \tfrac{1}{4}\pi.$$

Hence, $I=\frac{1}{2}\sqrt{\pi}$ or, in terms of the variable t.

$$\int_0^\infty e^{-t^2}\,dt=\tfrac{1}{2}\sqrt{\pi}. \tag{2}$$

Also, since e^{-t^2} is positive over the range $-\infty\leqslant t\leqslant 0$, then

$$\int_{-\infty}^\infty e^{-t^2}\,dt=\sqrt{\pi}. \tag{3}$$

(ii) In (2) put $t=x\sqrt{\alpha}$; then, if J denotes the resulting integral, we have

$$J\equiv\int_0^\infty e^{-\alpha x^2}\,dx=\frac{1}{2}\sqrt{\left(\frac{\pi}{\alpha}\right)}.$$

The integrand is continuous with respect to α and x in the range of integration; so also are the derivatives of the integrand with respect to α. Consider

$$\frac{\partial J}{\partial\alpha}\equiv-\int_0^\infty x^2\,e^{-\alpha x^2}\,dx$$

obtained by differentiating the integral J under the integral sign; the integral now obtained is uniformly convergent in the range. Hence on, differentiating $J\equiv\frac{1}{2}\sqrt{\left(\frac{\pi}{\alpha}\right)}$ with respect to α, we obtain

$$\int_0^\infty x^2\,e^{-\alpha x^2}\,dx=\frac{\sqrt{\pi}}{4}\,\frac{1}{\alpha^{\frac{3}{2}}}.$$

In general, if n is a positive integer, it is readily seen by successive applications of the procedure above that

$$\int_0^\infty x^{2n}\,e^{-\alpha x^2}\,dx=\frac{\sqrt{\pi}}{2}\,\frac{1.3.5.\ldots.2n-1}{2^n}\,\frac{1}{\alpha^{\frac{1}{2}(2n+1)}}.$$

Then, putting α equal to 1 and h^2 successively, we have the results:

$$\int_0^\infty x^{2n}\,e^{-x^2}\,dx=\frac{\sqrt{\pi}}{2}\,\frac{(2n)!}{2^{2n}n!} \tag{4}$$

and

$$\int_0^\infty x^{2n}\,e^{-h^2x^2}\,dx=\frac{\sqrt{\pi}}{2}\,\frac{(2n)!}{2^{2n}n!}\,\frac{1}{h^{2n+1}}. \tag{5}$$

Since the integrands are positive for $-\infty\leqslant x\leqslant 0$, the integrals between the limits $-\infty$ and ∞ have *twice* the values given on the right-hand sides of (4) and (5).

(iii) Let K denote the integral $\int_0^\infty x\,e^{-\alpha x^2}\,dx$. Then

$$K=\frac{1}{2\alpha}\int_0^\infty e^{-\alpha x^2}\,d(\alpha x^2)=\frac{1}{2\alpha}.$$

By the previous procedure

$$-\frac{\partial K}{\partial \alpha} \equiv \int_0^\infty x^3 e^{-\alpha x^2} dx = \frac{1}{2\alpha^2},$$

and, generally, if n is a positive integer, we easily deduce that

$$\int_0^\infty x^{2n+1} e^{-\alpha x^2} dx = \frac{1}{2} \frac{n!}{\alpha^{n+1}}.$$

We then derive, as in (ii),

$$\int_0^\infty x^{2n+1} e^{-x^2} dx = \tfrac{1}{2} n! \tag{6}$$

and

$$\int_0^\infty x^{2n+1} e^{-h^2 x^2} dx = \tfrac{1}{2} n! \frac{1}{h^{2n+2}}. \tag{7}$$

Since the integrands in (6) and (7) are negative for $-\infty \leqslant x \leqslant 0$, the integrals between the limits $-\infty$ and ∞ are *zero*.

(iv) *Integrals in terms of Gamma functions.* The Gamma function $\Gamma(m)$ is defined by

$$\Gamma(m) = \int_0^\infty t^{m-1} e^{-t} dt \quad (m > 0), \tag{8}$$

or, on writing $t = x^2$,

$$\Gamma(m) = 2\int_0^\infty x^{2m-1} e^{-x^2} dx. \tag{9}$$

In particular, from (8),

$$\Gamma(1) = \int_0^\infty e^{-t} dt = 1, \tag{10}$$

and, from (9) and (2),

$$\Gamma(\tfrac{1}{2}) \equiv 2\int_0^\infty e^{-x^2} dx = \sqrt{\pi}. \tag{11}$$

From (8), on integrating by parts,

$$\Gamma(m) = \left[\frac{1}{m} t^m e^{-t}\right]_0^\infty + \frac{1}{m}\int_0^\infty t^m e^{-t} dt.$$

Since $m > 0$, the integrated part vanishes at the lower limit, and, since $\lim_{t \to \infty} (t^m e^{-t}) = 0$, we have

$$\Gamma(m) = \frac{1}{m} \Gamma(m+1)$$

or

$$\Gamma(m+1) = m\Gamma(m). \tag{12}$$

If m is a positive integer, $\Gamma(m+1) = m!\,\Gamma(1)$ or, by (10),

$$\Gamma(m+1) = m! \tag{13}$$

or, from (9) on replacing m by $m+1$,

$$\int_0^\infty x^{2m+1} e^{-x^2} dx = \tfrac{1}{2} m!,$$

which is the same formula as (6); the formula (7) then follows.

Again, since $2m \equiv 2(m+\frac{1}{2})-1$, we have from (9)

$$\int_0^\infty t^{2m} e^{-t^2} dt = \tfrac{1}{2}\Gamma(m+\tfrac{1}{2}).$$

Now, by (12), $\Gamma(m+\frac{1}{2}) = (m-\frac{1}{2})\,\Gamma(m-\frac{1}{2})$.

If m is a positive integer then we obtain, by successive steps,

$$\begin{aligned}\Gamma(m+\tfrac{1}{2}) &= (m-\tfrac{1}{2})(m-\tfrac{3}{2})\ldots(\tfrac{1}{2})\,\Gamma(\tfrac{1}{2})\\ &= \frac{(2m-1)(2m-3)\ldots 1}{2^m}\frac{\sqrt{\pi}}{2}.\end{aligned}$$

Hence

$$\int_0^\infty t^{2m} e^{-t^2} dt = \frac{\sqrt{\pi}}{2}\frac{(2m)!}{2^{2m} m!}, \tag{14}$$

which is essentially the same formula as (4); the formula (5) then follows.

In Chapter 7 we shall have occasion to derive certain quantities in terms of the *Beta function* which is denoted by $B(m, n)$ and defined by

$$B(m,n) = \int_0^1 x^{m-1}(1-x)^{n-1} dx = 2\int_0^{\frac{1}{2}\pi} \cos^{2m-1}\theta \sin^{2n-1}\theta\, d\theta, \tag{15}$$

in which $m>0$ and $n>0$. The Beta function can be expressed in terms of Gamma functions as follows: from (9)

$$\Gamma(m)\,\Gamma(n) = 4\int_0^\infty\int_0^\infty \xi^{2m-1} e^{-\xi^2} \eta^{2n-1} e^{-\eta^2} d\xi\, d\eta;$$

if $\xi = r\cos\theta$ and $\eta = r\sin\theta$ so that $d\xi\, d\eta = r\, dr\, d\theta$, the right-hand side becomes

$$4\int_0^\infty r^{2m+2n-1} e^{-r^2} dr \int_0^{\frac{1}{2}\pi} \cos^{2m-1}\theta \sin^{2n-1}\theta\, d\theta.$$

Hence

$$B(m,n) = \frac{\Gamma(m)\,\Gamma(n)}{\Gamma(m+n)}. \tag{16}$$

(v) *Summary of integrals.* In terms of the variable t we have the following results, m being a positive integer:

$$\int_0^\infty e^{-t^2}\,dt=\tfrac{1}{2}\sqrt{\pi},\quad \int_{-\infty}^\infty e^{-t^2}\,dt=\sqrt{\pi}; \tag{17}$$

$$\int_0^\infty t^{2m}\,e^{-t^2}\,dt\equiv\tfrac{1}{2}\Gamma(m+\tfrac{1}{2})=\frac{\sqrt{\pi}}{2}\,\frac{(2m)!}{2^{2m}m!}; \tag{18}$$

$$\int_0^\infty t^{2m+1}\,e^{-t^2}\,dt\equiv\tfrac{1}{2}\Gamma(m+1)=\tfrac{1}{2}m!. \tag{19}$$

When the limits of integration are $-\infty$ *and* $+\infty$ the definite integral corresponding to (18) has twice the value given on the right-hand side of (18), and the definite integral corresponding to (19) is zero.

The corresponding integrals involving h are obtained by putting $t=hx$ in the previous results; we have then:

$$\int_0^\infty e^{-h^2x^2}\,dx=\frac{\sqrt{\pi}}{2h},\quad \int_{-\infty}^\infty e^{-h^2x^2}\,dx=\frac{\sqrt{\pi}}{h}; \tag{20}$$

$$\int_0^\infty x^{2m}\,e^{-h^2x^2}\,dx=\frac{\sqrt{\pi}}{2h}\,\frac{(2m)!}{(2h)^{2m}\,m!}; \tag{21}$$

$$\int_0^\infty x^{2m+1}\,e^{-h^2x^2}\,dx=\frac{1}{2h^2}\,\frac{m!}{h^{2m}}. \tag{22}$$

The values of these h integrals when the limits of integration are $-\infty$ and $+\infty$ are obtained as in the case of (18) and (19).

1·15. Properties of the normal distribution function

For convenience we denote by x (instead of ξ, as in § 1·13) the deviation of the variable from the mean so that the normal function is given by

$$y\equiv f(x)=\frac{h}{\sqrt{\pi}}\,e^{-h^2x^2}, \tag{1}$$

and the relative frequency for values of the deviation between x and $x+dx$ is $f(x)\,dx$ or $\dfrac{h}{\sqrt{\pi}}e^{-h^2x^2}\,dx$.

(i) *The normal curve.* The curve given by (1) and shown in Fig. 6 is symmetrical about the y-axis, and it is obvious that the mean, $\bar{x}$, is zero; also, when $x=0$, then $y\equiv OA=h/\sqrt{\pi}$, and when $x=\pm\infty$, then $y=0$ so that the x-axis is an asymptote. We refer to the curve as the *normal curve.*

The area under the curve is

$$\int_{-\infty}^\infty y\,dx\equiv\frac{h}{\sqrt{\pi}}\int_{-\infty}^\infty e^{-h^2x^2}\,dx=1,$$

by 1·14 (20). Thus, the total relative frequency is 1, as we would expect.

From (1),
$$\frac{dy}{dx} = -\frac{2h^3}{\sqrt{\pi}} x\, e^{-h^2x^2}$$
and
$$\frac{d^2y}{dx^2} = -\frac{2h^3}{\sqrt{\pi}} \{1 - 2h^2x^2\}\, e^{-h^2x^2}.$$

In this last formula the expression on the right-hand side vanishes if $x^2 = 1/(2h^2)$; hence there are points of inflexion in Fig. 6 at B and B_1 whose abscissae are $\pm 1/(h\sqrt{2})$.

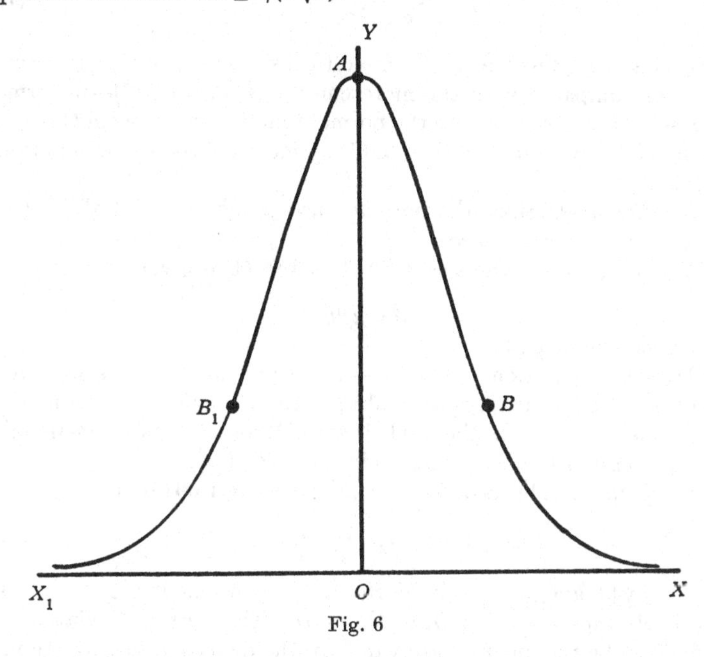

Fig. 6

(ii) *Principal moments.* Since the normal curve is symmetrical about the y-axis ($x=0$) corresponding to the mean of the deviations in the range $-\infty$ to $+\infty$, the principal moment of r-th order is given by

$$\mu_r = \frac{h}{\sqrt{\pi}} \int_{-\infty}^{\infty} x^r\, e^{-h^2x^2}\, dx. \tag{2}$$

From 1·14 (22) and (21) it follows that

$$\mu_1 = \mu_3 = \ldots = 0, \tag{3}$$

$$\mu_2 = \frac{1}{2h^2}, \tag{4}$$

and

$$\mu_4 = \frac{3}{4h^4}. \tag{5}$$

(iii) *Variance and standard deviation.* By definition, μ_2 is the variance σ^2 and the standard deviation is σ; hence, by (4),

$$\sigma^2 = \frac{1}{2h^2} \quad \text{or} \quad h^2 = \frac{1}{2\sigma^2}. \tag{6}$$

In statistical theory the normal function $f(x)$ is expressed in terms of σ so that, by (6),

$$f(x) = \frac{1}{\sigma\sqrt{(2\pi)}} e^{-x^2/2\sigma^2}. \tag{7}$$

Having regard to the greater analytical simplicity of (1)—in terms of h—as compared with the more cumbrous form of (7)—in terms of σ—we shall generally use the normal function in the form (1) and, when necessary, translate (1) into (7) by means of the second relation in (6).

(iv) *Skewness.* Since the normal curve is symmetrical about the y-axis, its skewness is zero.

Formally, the skewness as defined by 1·09 (1) is given by

$$\beta_1^{\frac{1}{2}} = \mu_3/\sigma^3$$

which vanishes by (3).

Pearson's definition of skewness is (mean − mode)/σ; as we have mentioned the mean, $\bar{x}$, is zero; also, since the mode is the value of x when y is a maximum, the mode is zero. Hence, formally, Pearson's definition gives zero-skewness, as in the case of $\beta_1^{\frac{1}{2}}$.

(v) *Kurtosis.* This is defined by β_2, given by 1·09 (5), namely,

$$\beta_2 = \mu_4/\mu_2^2 = \mu_4/\sigma^4.$$

Hence, by (4) and (5), $\beta_2 = 3$; this result was mentioned earlier in § 1·09.

(vi) *Average deviation from the mean.* The *average* deviation is defined to be the mean of the values of the deviation taken without regard to sign. Since the normal curve is symmetrical about the y-axis we need consider only the positive deviations, that is, those between 0 and $+\infty$. If, as before, we suppose that the total number of deviations is N, then there will be $\frac{1}{2}N$ positive deviations and the number of deviations between x and $x+dx$ will be $Nf(x)\,dx$. Hence, if η denotes the average deviation,

$$\tfrac{1}{2}N\eta = N\int_0^\infty x f(x)\,dx = N\frac{h}{\sqrt{\pi}}\int_0^\infty x\,e^{-h^2x^2}\,dx.$$

By 1·14 (22), the value of the integral is $1/(2h^2)$; hence

$$\eta = \frac{1}{h\sqrt{\pi}}, \tag{8}$$

or, in terms of the standard deviation,

$$\eta=\sigma\sqrt{\left(\frac{2}{\pi}\right)}. \tag{9}$$

Thus $$\eta:\sigma=0\cdot798.$$

(vii) *The error function.* In the next chapters the normal function will be met with, *errors* taking the place of deviations. The results which have been derived in § 1·14 and in the present section form the basis of the subsequent investigations. An important function is the *error function*, which is closely related to the normal function and its integral with some precise upper limit. As in (vi), the frequency of deviations from the mean between x and $x+dx$ is $Nf(x)\,dx$. Hence the total frequency of deviations between 0 and ξ is

$$N\frac{h}{\sqrt{\pi}}\int_0^{\xi}e^{-h^2x^2}\,dx \quad \text{or} \quad \frac{N}{\sqrt{\pi}}\int_0^{h\xi}e^{-t^2}\,dt,$$

where $t=hx$.

The error function, denoted by $\operatorname{erf}(t)$, is defined by

$$\operatorname{erf}(t)=\frac{2}{\sqrt{\pi}}\int_0^{t}e^{-t^2}\,dt. \tag{10}$$

Tables of this function are given in Appendix 1.

Thus, when h is known, the total number of deviations from the mean between 0 and an assigned value of ξ is given by

$$\tfrac{1}{2}N\operatorname{erf}(h\xi).$$

We illustrate the use of the error function in the following section.

1·16. Example of a normal distribution

We consider the statistical distribution of the heights of 8505 men of medical grade II,† which proves, as a result of subsequent calculations, to be represented very closely by a normal curve (Fig. 7). In Table 5 the class interval, c, is 2 in. and the first column gives the middle height for each of the class intervals—that is, the abscissa of the corresponding point on the frequency polygon; the second column gives the frequencies, f.

If x denotes height in inches we can simplify the calculations in the usual way by writing

$$x=67+cu\equiv a+cu. \tag{1}$$

The next columns show the various details of the subsequent calculations.

† W. J. Martin, *Medical Research Council Memoranda*, no. 20 (H.M.S.O. 1949); C. G. Lambe, *Elements of Statistics* (Longmans, 1952), p. 3.

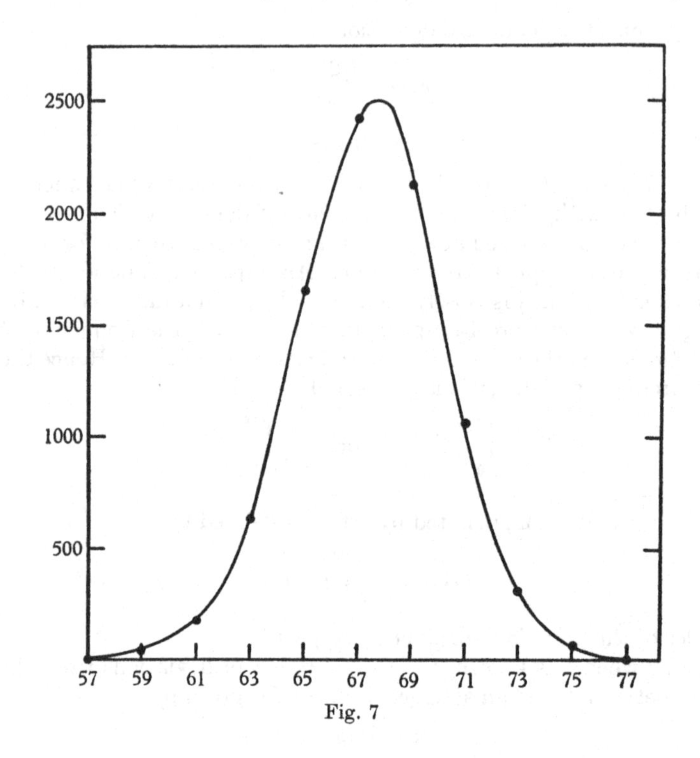

Fig. 7

Table 5. *Heights of* 8505 *men*

Middle height	f	u	fu	fu^2	fu^3	fu^4	$\|\eta\|$	$f\|\eta\|$
57	3	−5	− 15	75	− 375	1,875	10·47	31
59	30	−4	− 120	480	− 1,920	7,680	8·47	254
61	158	−3	− 474	1,422	− 4,266	12,798	6·47	1,022
63	639	−2	−1,278	2,556	− 5,112	10,224	4·47	2,856
65	1,648	−1	−1,648	1,648	− 1,648	1,648	2·47	4,071
67	2,439	0	**−3,535**	0	**−13,321**	0	0·47	1,146
69	2,124	+1	+2,124	2,124	+ 2,124	2,124	1·53	3,250
71	1,065	+2	+2,130	4,260	+ 8,520	17,040	3·53	3,759
73	327	+3	+ 981	2,943	+ 8,829	26,487	5·53	1,808
75	66	+4	+ 264	1,056	+ 4,224	16,896	7·53	497
77	6	+5	+ 30	150	+ 750	3,750	9·53	57
	8,505		**+5,529**	**16,714**	**+24,447**	**100,522**		**18,751**

Summary: $N=8{,}505$; $\Sigma fu=1{,}994$; $\Sigma fu^2=16{,}714$; $\Sigma fu^3=11{,}126$; $\Sigma fu^4=100{,}522$; $\Sigma f|\eta|=18{,}751$.

If the actual distribution is a true normal distribution, then the numerical values of the principal moments and of the average deviation should lead to the same value of h, for each of these theoretical quantities depends only on h. We test the statistics in this way, using the summations at the foot of Table 5 to calculate the principal moments and the average error. With the usual notation we have the following.

(i) *The mean*
$$\bar{u} = \frac{1}{N}\Sigma fu = \tfrac{1994}{8505} = 0{\cdot}2344.$$

Hence, from (1),
$$\bar{x} = 67{\cdot}469.$$

(ii)
$$\mu_2(a) = c^2\mu_2'(a) = 7{\cdot}861.$$

(iii)
$$\mu_3(a) = c^3\mu_3'(a) = 10{\cdot}465.$$

(iv)
$$\mu_4(a) = c^4\mu_4'(a) = 189{\cdot}106.$$

(v) $\bar{\xi} \equiv \bar{x} - a = 0{\cdot}469$; $\bar{\xi}^2 = 0{\cdot}220$; $\bar{\xi}^3 = 0{\cdot}103$; $\bar{\xi}^4 = 0{\cdot}048$.

(vi) From 1·15 (6),
$$\mu_2 \equiv \mu^2 = \frac{1}{2h^2} = \mu_2(a) - \bar{\xi}^2 = 7{\cdot}661.$$

Hence
$$h = 0{\cdot}256.$$

(vii) From 1·06 (16),
$$\mu_4 = \mu_4(a) - 4\bar{\xi}\mu_3(a) + 6\bar{\xi}^2\mu_2(a) - 3\bar{\xi}^4$$
$$= 179{\cdot}71.$$

But, from 1·15 (5)
$$\mu_4 = \frac{3}{4h^4}.$$

Hence
$$h = 0{\cdot}254.$$

(viii) The numerical values of the deviations from the mean are denoted by $|\eta|$ in the table. The average deviation, η, is given by
$$\eta = \frac{1}{N}\Sigma f|\eta| = \frac{18{,}751}{N} = 2{\cdot}205.$$

But, from 1·15 (8),
$$\eta = \frac{1}{h\sqrt{\pi}}.$$

Hence
$$h = 0{\cdot}256.$$

(ix) The values of μ_2, μ_4 and η give values of h which are all very much alike, and we conclude that the statistical distribution is, to a high degree of approximation, a normal distribution for which we can take $h=0{\cdot}255$. This value is also seen to be in quite good accordance with the maximum ordinate of the curve (Fig. 7) for the maximum ordinate of the relative frequency distribution is $h/\sqrt{\pi}$ which is to be equated—from the curve—to $\frac{1}{cN}2500$. Thus, approximately, $h=0{\cdot}260$—a value that is liable to some uncertainty since it depends on smoothing the frequency polygon in the neighbourhood of the maximum.

(x) We compare the number of men with heights between x_1 and x_2 with the theoretical number. From Table 5 the number of men between 68 and 76 in. is 2124+1065+327+66 or 3582. If ξ_1 and ξ_2 are the corresponding deviations from the mean $\bar{x}$ $(=67{\cdot}469)$, then $\xi_1=0{\cdot}531$ and $\xi_2=8{\cdot}531$. The theoretical number, n, is then given by

$$n=\frac{Nh}{\sqrt{\pi}}\int_{\xi_1}^{\xi_2} e^{-h^2x^2}\,dx,$$

or

$$n=\frac{N}{\sqrt{\pi}}\int_{h\xi_1}^{h\xi_2} e^{-t^2}\,dt=\frac{N}{2}[\operatorname{erf}(h\xi_2)-\operatorname{erf}(h\xi_1)].$$

If $h=0{\cdot}255$, then $h\xi_1=0{\cdot}1354$ and $h\xi_2=2{\cdot}1754$. Then, from the tables of the error function (Appendix 1),

$$n=\frac{8505}{2}[0{\cdot}99790-0{\cdot}15185]$$

$$=3598,$$

which may be regarded as being in close agreement with the total of 3582 obtained by direct counting.

1·17. Sheppard's corrections

In calculating the principal moments for a statistical distribution as in the previous section it is to be remembered that we assume that all the individuals, within the class interval, c, have the *same* value of the characteristic, x. The actual distribution is, however, continuous and Sheppard's corrections† enable us, *in certain circumstances*, to derive the principal moments for the continuous or true distribution.

If $\mu_1, \mu_2, \ldots$ are the principal moments as calculated and $(\mu_1), (\mu_2), \ldots$

† Derived in § 8·03; see also E. T. Whittaker and G. Robinson, *The Calculus of Observations* (Blackie, 1924), p. 194.

are the corrected principal moments (which now refer to the continuous distribution), then it is found (see § 8·03) that

$$(\mu_1)=\mu_1=\bar{x};\quad (\mu_2)=\mu_2-\tfrac{1}{12}c^2;$$

$$(\mu_3)=\mu_3;\quad (\mu_4)=\mu_4-\tfrac{1}{2}c^2\mu_2+\tfrac{7}{240}c^4.$$

These formulae embody Sheppard's corrections.

The circumstances in which the application of the formulae is justifiable are stated in § 8·03. It need only be mentioned here that the formulae can be legitimately applied in the case of a normal distribution such as that in § 1·16.

Consider the statistics in the previous section. Then, with $c=2$, it is easily found that

$$(\mu_2)=7{\cdot}328 \quad\text{and}\quad (\mu_4)=164{\cdot}86.$$

Further, $$(\mu_2)=\frac{1}{2h^2}\quad\text{and}\quad(\mu_4)=\frac{3}{4h^4},$$

from which we have $h=0{\cdot}261$ and $h=0{\cdot}260$ respectively.

Referring to § 1·16 (x), we find that the number, n, of men with heights between 68 and 76 in. is given, with $\xi_1=0{\cdot}531$ and $\xi_2=8{\cdot}531$ as before and with $h=0{\cdot}260$, by

$$n=\tfrac{1}{2}N[\operatorname{erf}(2{\cdot}21806)-\operatorname{erf}(0{\cdot}13806)]$$

$$=\tfrac{1}{2}.8805[0{\cdot}99829-0{\cdot}15480]$$

$$=3587.$$

This result is in much closer agreement with the counted number (3582) than the number (3598) derived from the uncorrected moments.

We conclude that the distribution has all the characteristics of a normal distribution.

1·18. Note on the evaluation of erf *t*

It is convenient to deal with the function $\phi(t)$ defined by

$$\phi(t)=\int_0^t e^{-t^2}\,dt, \tag{1}$$

so that $$\operatorname{erf}(t)=\frac{2}{\sqrt{\pi}}\phi(t). \tag{2}$$

Since $$e^{-t^2}=\sum_0\frac{(-1)^n t^{2n}}{n!},$$

we obtain from (1) $\phi(t) = \sum_0 \frac{(-1)^n t^{2n+1}}{(2n+1)\,n!}$

$$= t - \frac{t^3}{3.1!} + \frac{t^5}{5.2!} - \dots,$$

which converges for all values of t; this series is useful when $t<1$ when it converges rapidly.

When $t>1$, write

$$\phi(t) = \int_0^\infty e^{-t^2}\,dt - \int_t^\infty e^{-t^2}\,dt \equiv \tfrac{1}{2}\sqrt{\pi} - \phi_1(t).$$

Now
$$\int e^{-t^2}\,dt = -\frac{1}{2t}e^{-t^2} - \frac{1}{2}\int \frac{1}{t^2}e^{-t^2}\,dt$$

$$= -\frac{1}{2t}e^{-t^2} + \frac{1}{4t^3}e^{-t^2} + \frac{1.3}{2.2}\int \frac{1}{t^4}e^{-t^2}\,dt.$$

Proceeding in this way and inserting the limits we find that

$$e^{t^2}\phi_1(t) = \frac{1}{2t} - \frac{1}{2}\frac{1}{2t^3} + \frac{1.3}{2^2}\frac{1}{2t^5} + \dots + (-1)^n \frac{1.3.\dots.2n-1}{2^n}\frac{1}{2t^{2n+1}}$$

$$-(-1)^n \frac{1.3.\dots.2n-1}{2^n}\frac{2n+1}{2}e^{t^2}\int_t^\infty \frac{e^{-t^2}}{t^{2n+2}}\,dt. \quad (3)$$

Now
$$e^{t^2}\int_t^\infty \frac{e^{-t^2}\,dt}{t^{2n+2}} < \int_t^\infty \frac{dt}{t^{2n+2}} < \frac{1}{2n+1}\frac{1}{t^{2n+1}}.$$

Hence, if T_n denotes the last integrated term in (3) and R_n the final term, then $|R_n| < |T_n|$. We have then the asymptotic series

$$\phi_1(t) = \frac{e^{-t^2}}{2t}\left[1 - \frac{1}{2}\frac{1}{t^2} + \frac{1.3}{2^2}\frac{1}{t^4} - \frac{1.3.5}{2^3}\frac{1}{t^6} + \dots\right],$$

which enables us to calculate the value of $\phi_1(t)$ with an error less than the term T_n.

Thus $\phi(t)$ is obtained and, finally, erf(t).

CHAPTER 2

ERRORS OF OBSERVATION AND THE PRINCIPLE OF LEAST SQUARES

2·01. Introduction

In this chapter we deal with the statistics relating to observations or measurements of a particular quantity, obtained by means of instruments designed to provide the greatest possible precision. For example, the velocity of light, c, has been measured by many physicists employing a variety of experimental methods (see Table 11, p. 105). For a particular experimenter the statistical raw material consists of perhaps twenty independent determinations of this fundamental physical quantity; if these measures are equally reliable then, according to principles to be discussed later, he will take the mean, c_1, to represent his final result. Other experimenters will similarly obtain results $c_2, c_3, \ldots$. In general, these results, although all different, will be found to cluster closely around the mean, $\bar{c}$. Now, c has a definite value, and so a particular result, c_1, has associated with it the error $c_1 - c$, which is unknown, since the *precise* value of c is unknown. The problem is to derive that value, c_0, which best represents the aggregate of results and to assign an estimate of the accuracy of c_0 according to some criterion. The procedure to be followed is based on the study of the nature of errors in general and the way they arise in any series of observations or measurements. In the next four sections we describe the several kinds of errors† associated with observations.

2·02. Instrumental errors

To discuss the matter in some detail we take a typical example from observational astronomy—the measurement of the right ascension, α, of a star.‡ In this connexion the fundamental instrument is the 'meridian circle' or 'transit instrument', which consists primarily of a telescope set up, ideally, with its rotational axis horizontal and oriented in the east-west direction. Mounted symmetrically

† We exclude gross errors—for example, in reading an angular scale with an error of 1°—and serious mistakes in computation, all of which can, presumably, be detected on scrutinizing the results.

‡ Right ascension is analogous to terrestrial longitude measured, however, westward from a particular reference point from 0h. to 24h., equivalent to the range 0° to 360°.

in the focal plane of the object-glass, there is a group of seven spider's webs or 'vertical wires' as shown in Fig. 8; ideally, the optical axis of the object-glass meets the central wire OP midway between O and P. When the telescope is rotated about its horizontal axis, OP will sweep out—under these ideal conditions—the meridian of longitude on which the instrument is situated, specified more particularly as the longitude of the central wire. To make an observation of a particular star, the telescope is pointed in the appropriate direction (known, with sufficient accuracy, from the latitude of the observatory and the approximate declination of the star) and then clamped; for a minute or two before the star's meridian passage (or transit) and for a minute or two after, the star is seen in the field of view moving along some such path as XY by reason of the east-west diurnal motion of the celestial sphere.

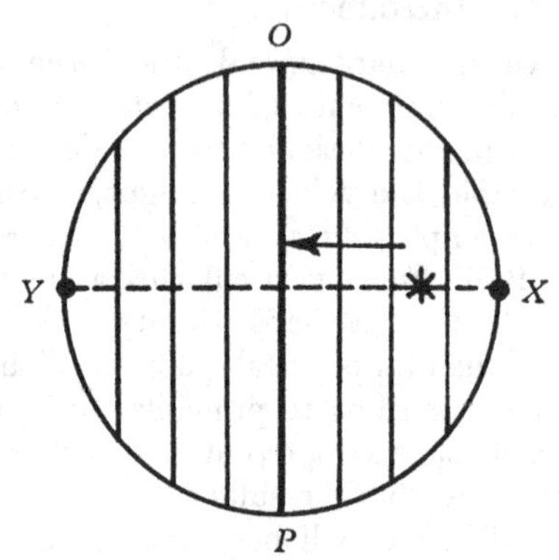

Fig. 8

For the observations in view, a second instrument is required; this is an accurate clock keeping, ideally, the sidereal time corresponding to the longitude of the telescope. In the old-time observations the clock was situated close to the telescope within view of the observer so that he could note the beginning of a particular minute and hear the beats of the subsequent seconds; in modern practice the clock is kept in a thermostatically controlled chamber and is electrically connected with a chronograph and subsidiary apparatus on the telescope.

The observations consist in recording the clock times at which the image of the star appears to be coincident with each of the vertical wires. The mean, T, of the times of transit over the seven wires is taken to be the instant of meridian passage. In these ideal circumstances the right ascension, α, of the star is given† by

$$\alpha = T. \tag{1}$$

But no instrument is perfect, and we have to recognize that the telescope axis is *not* accurately in the east-west direction and that it is *not* accurately level and, further, that the optical axis of the object-glass does not pass exactly through OP in Fig. 8; these deviations

† We omit a small correction, which can be accurately calculated, due to 'diurnal aberration'; also, the precise definition of α with respect to the appropriate reference point and reference plane need not be described.

from ideal conditions are respectively the azimuth error, a, the level error, b, and the collimation error, c. Further, the setting of the sidereal clock is not likely to be exact, for it may be a little fast or a little slow on the 'correct' time; the error, denoted algebraically by ΔT, is called the clock error, so that, if T is the observed time of transit, the correct time is $T+\Delta T$.

It is shown in books on spherical astronomy† that with the restriction mentioned—relating to diurnal aberration—the equation (1) is replaced by

$$\alpha = T + \Delta T + aA + bB + cC, \tag{2}$$

in which A, B and C are known constants for the observatory concerned and for the particular star observed.

It is thus clear that, to determine α with the greatest degree of precision, the values of ΔT and the instrumental errors a, b and c must be known as accurately as possible. It is sufficient to state that these values are derived by means of subsidiary observations and experiments and are themselves subject to observational errors—and also to change, as is simply seen, for example, in connexion with ΔT which is influenced by the 'clock rate', this latter being itself the subject of special investigations.

2·03. Personal equation

A second type of error arises from the psychological reactions of the observer. For example, when a chronograph is used in conjunction with a sidereal clock to record the time of transit over a wire, the observer presses a key, connected with the chronograph in an electrical circuit, when he adjudges the image of the star to be in coincidence with a vertical wire; his peculiarity may be that he is anxious not to be 'late' in pressing the key and consequently he is apt to record the observation a little before the star is actually 'on the wire'. The resulting error is known as the *personal equation* of the observer, the evaluation of which in the past has generally been a matter of considerable difficulty; with modern instruments and devices the personal equation in meridian observations has been greatly reduced. In other kinds of observations and measurements personal equation may still be a source of embarrassment, requiring special methods for its detection and elucidation.

The history of personal equation goes back to near the end of the eighteenth century when the 'eye and ear' method‡ of observing

† E.g. W. M. Smart, *Spherical Astronomy*, 4th ed. (Cambridge University Press, 1956), p. 80.

‡ In this method the observer, listening to the beats of the sidereal clock nearby, estimated to a tenth of a second the time of transit of the star across a wire; hence the name of the method.

transits was practised. In 1796 Maskelyne, fifth Astronomer Royal, wrote:†

> My assistant, Mr David Kinnebrook, who had observed transits of stars and planets very well in agreement with me all the year 1794 and for a great part of [1795], began from the beginning of August last to set them down half a second later than he should do according to my observations; and, in January [1796] he increased his error to eight-tenths of a second. As he had unfortunately continued a considerable time in this error before I noticed it, and did not seem to me likely ever to get over it and return to the right‡ method of observing, therefore, although with reluctance, as he was a diligent and useful assistant to me in other respects, I parted with him.

It was fully a quarter of a century later that the existence of personal equation was firmly established by Bessel—his attention had been drawn then to the Maskelyne-Kinnebrook episode—and many investigations were undertaken for the study of its incidence and peculiarities as it affected different observers. The unfortunate Kinnebrook was undoubtedly a victim to a change in the personal equation (then, of course, unknown) of either himself, or of Maskelyne, or of both; it is certain that, in more enlightened times, he would not have suffered the injustice of dismissal.

2·04. Systematic error

A third kind of error is the systematic error which may be constant or periodic in character. Two illustrations of the constant systematic error are as follows. If the true zero of a sextant scale has not been accurately determined by the special observations made for this purpose so that the 'index error' is somewhat erroneous, a residual error remains which affects all readings by the same constant amount. A second example relates to a skilful marksman firing at a target; if there is a cross-wind blowing of which he is unaware and if he is not informed where each shot hits the target, the mean point of the pattern of hits will be somewhat displaced from the centre of the target, and this deviation is a constant systematic error affecting all his shots under the circumstances stated.

Again, some classes of observations are affected by meteorological conditions which are approximately reproduced in cycles of a year; the errors introduced are thus periodic in character.

The observer, then, should always be on the look-out for possible systematic error and, if this is confirmed, he should devise means

† *Greenwich Observations*, **3**; see also E. C. Sanford, *Amer. J. Psychol.* **2**, 1. 271, 403 (1889).

‡ Maskelyne averred that Kinnebrook had fallen 'into some irregular and confused method of his own'.

for its eradication. He may even be led to make a fundamental discovery. The classical example is Bradley's discovery of the phenomenon of aberration. Attempting to derive the parallax of the second-magnitude star γ Draconis, Bradley measured, with a fixed instrument, the star's declination at transit as opportunity permitted; he found that his measures revealed a periodic variation with the period of a year, but 90° out of phase with the anticipated periodic variation associated with parallax displacements and of a magnitude far in excess of any possible observational error. He was led to reflect on how such a periodicity could arise and, eventually, to explain the phenomenon in terms of what we now call† 'annual aberration'. A few years later he discovered, by similar methods, the phenomenon of nutation.

2·05. Accidental error

A fourth kind of error, with which we shall be mainly concerned, is accidental error. The accuracy of an observation with the meridian circle, and, indeed, with most astronomical instruments, is dependent to a large extent on atmospherical conditions. If the 'seeing' is perfect, the star moves across the field of view at a uniform rate and the observation of the time of transit over a wire would be expected —other considerations apart—to be highly accurate. But, in practice, the conditions are not quite as depicted; owing to atmospherical disturbances the star moves across the field of view at a slightly irregular rate and along a slightly irregular course; accordingly, the recording of the time of transit over a wire is subject to uncertainty, the error which expresses this uncertainty being as likely to be positive or negative according to the vagaries of the atmosphere throughout the path of the ray. Such an error is said to be *accidental* and is the result, in general, of fortuitous and unpredictable irregularities in observational and instrumental conditions. In particular, the magnitude of the accidental error is likely to be greater for poor 'seeing' than for good.

As previously mentioned the mean time, T, of the times T_i ($i = 1, 2, \ldots, 7$), over the individual wires is taken to be the time of meridian transit. Since each T_i is susceptible of an accidental error ϵ_i which may be positive or negative, the mean T is, on general grounds, likely to be a more accurate result than for any individual T_i; this, is, of course, the reason why the observations are made over seven wires (or some similar number).

† In seconds of arc the constant of aberration when the eccentricity of the earth's orbit is neglected is v/c cosec $1''$ or about $20''\cdot5$, where v is the earth's velocity, and c the velocity of light.

Again, in the 'eye and ear' method of observing transits, current until about a century ago, the time of transit over each wire was estimated to a tenth of a second; thus, if the object of the observer over a series of observations was to derive a final result in terms of hundredths of a second, an error up to five-hundredths of a second in magnitude would appear in each estimation; such errors, clearly, could be positive or negative and are to be classed as accidental errors. Similar remarks are applicable to the readings of any instrument when the final accuracy aimed at is greater than the accuracy with which a single reading can be made.

The sources of accidental errors are many and varied—we have mentioned only a few—and the determination of, say, the right ascension of a star is subject to an unknown accidental error ϵ which is made up of a large number of individual accidental errors—or 'elementary errors' as we may describe them—which can be positive or negative. The error, ϵ, of the final determination is also accidental in character, for it depends on the unknown way in which the elementary errors are combined.

2·06. Summary of errors in general

(i) The effects of *instrumental errors*, such as ΔT, a, b and c in transit observations, can be substantially removed from the observations by means of special investigations. But even when ΔT, a, b and c have been determined with the greatest possible accuracy, residual errors and errors resulting later from small changes in these quantities still remain which have, over a series of observations, the general characteristics of accidental errors.

(ii) The possibility of the existence of personal equation and of systematic error must always be borne in mind, and these, if established, should be made a special study. Again, residual errors may be expected to be of the accidental type.

(iii) There are, finally, the accidental errors (including those already mentioned) which arise in a fortuitous way. Their presence in the observations must be recognized, although their effects cannot be directly or specifically allowed for, as in the case of instrumental errors.

2·07. Combination of observations

We continue to take as an illustration the determination of the right ascension, α, of a star; we assume that, from a series of observations made by one or more observers, the values α_i $(i=1,2,\ldots,n)$ are obtained. It is also assumed that each observation has been corrected as far as possible for all errors other than accidental errors;

thus, associated with α_i is an accidental error ϵ_i of which the magnitude and sign are unknown.

The problem confronting us in brief, is (i) to combine the n results α_i according to some acceptable precept so as to determine the value of α which best represents the aggregate of observations, and (ii) to assess the precision of this value according to a specific criterion.

Occasionally we are concerned only with (ii) when the value of the quantity to be determined is accurately known; one simple example is the experimental determination of the value of π from the ratio of the measure of the circumference of a circle to the measure of its diameter; another example is the experimental determination of the value of g when, this value being known, the object is to test the capabilities of a new instrument or perhaps to assess the experimental skill of a class of students using the apparatus familiar in the laboratory such as the simple pendulum or Attwood's machine.

In the following pages our principal study is the theory of errors —the general problem defined by (i) and (ii) above—treated in the remainder of this chapter by elementary methods and later according to the concepts of probability. It must be emphasized that the theory of errors, unless otherwise stated explicitly, is concerned with accidental errors only.

2·08. Observational equations in one unknown

Let x denote the *true* value of the quantity to be determined, and x_i $(i=1,2,\ldots,n)$ the observed or measured values. For simplicity it may be supposed in this chapter that the n observations have been carefully made by the same observer working in comparable conditions throughout; the observations may then be deemed to be equally reliable or, in statistical language, of *equal weight.*

If ϵ_i is the error of x_i, then

$$x_i - x = \epsilon_i.$$

This is an accurate equation in which, however, only one element, x_i, is actually known.

The equation $$x_i - x = 0$$

is regarded symbolically as the general *observational equation*, corresponding to the observed value x_i; as such it implies the existence of an error ϵ_i.

2·09. The postulate of the arithmetic mean

From the previous section the exact equation associated with each of n observations is of the form

$$x_i - x = \epsilon_i. \tag{1}$$

There are n such equations containing $(n+1)$ unknowns, namely, x and the n errors ϵ_i; consequently, we are unable to derive the *true* value, x, from these equations. To derive *some* value of x we must have another equation, and to derive an acceptable value of x this equation should be equivalent to a reasonable hypothesis relating to the accidental errors ϵ_i.

Any *arbitrary* function of the errors can be written as

$$\phi(\epsilon_1, \epsilon_2, \ldots, \epsilon_n)=0; \tag{2}$$

accordingly, if a particular form of ϕ is assumed, the n equations (1) and the equation (2) can be solved to produce a definite value of x. But it must be noted that the value of x obtained in this way depends on the hypothesis introduced, represented by (2), and is not necessarily the *true* value of the unknown.

A simple arbitrary relation of the form (2) is

$$\sum_{i=1}^{n} \epsilon_i \equiv \epsilon_1+\epsilon_2+\ldots+\epsilon_n=0. \tag{3}$$

The solution of the n equations (1) and the equation (3) is easily achieved. Add the n equations (1); then†

$$\Sigma x_i - nx = \Sigma\epsilon_i = 0, \quad \text{by (3)},$$

or

$$x=\frac{1}{n}\Sigma x_i. \tag{4}$$

We denote by a the arithmetic mean of the n measures x_i, a simpler nomenclature than the usual $\bar{x}$. Then (4) becomes

$$x=a.$$

Thus, the true value of the unknown is the arithmetic mean *provided that the arbitrary relation* (3) *is in fact a true relation.*

As regards the 'reasonableness' of the hypothesis embodied in (3) it is to be remarked that of the accidental errors $\epsilon_1, \ldots, \epsilon_n$, some are positive and some are negative; this is the nature of accidental errors. If $n=10$, say, it would hardly be expected that $\Sigma\epsilon_i$ would be exactly zero; on the other hand, it would be expected that $\frac{1}{n}\Sigma\epsilon_i$ would be very much smaller than the *average* error, taken without regard to sign and given by $\frac{1}{n}\Sigma|\epsilon_i|$; accordingly, the arithmetic mean would

† As in Chapter 1 the limits of summation will be omitted for simplicity when no confusion is likely to be caused.

be regarded ordinarily as a good *approximation* to the value of the unknown.

In many classes of observations in which it is believed that accidental errors alone are present, the arithmetic mean is generally found to give a satisfactory and consistent value of the unknown quantity x; thus, if several series of observations are undertaken by skilled observers, each series consisting of a number of individual observations, the arithmetic means for the several series are found to differ very little amongst themselves.

Instead of introducing a hypothesis such as (3) we take as a basic postulate that the *best* value of the unknown is the arithmetic mean; this is known as the *postulate of the arithmetic mean.*

As we shall see in Chapter 3 the postulate of the arithmetic mean is fundamental in Gauss's investigation of the normal law of errors based on the principles of probability.

2·10. Residuals

The difference between an observed value x_i and any assigned value, ξ, of x is called a *residual*, described more conveniently in the present connexion as a ξ residual. Thus, corresponding to x_i we have the ξ residual denoted by V_i and defined by

$$V_i = x_i - \xi. \tag{1}$$

If ξ is taken to be the arithmetic mean, a, of the n observed quantities $x_1, x_2, \ldots, x_n$, the residual for x_i is denoted by v_i and defined by

$$v_i = x_i - a. \tag{2}$$

The residual v_i is equivalent to the deviation from the arithmetic mean as used in § 1·06.

When the term 'residual' is used in this chapter without further specification, the sense in which it is described is by means of (2), that is, with reference to the arithmetic mean.

2·11. The principle of least squares

As before, a denotes the arithmetic mean of the measures x_i $(i = 1, 2, \ldots, n)$ of a quantity whose true value, x, is unknown; then

$$a = \frac{1}{n} \Sigma x_i. \tag{1}$$

Let ξ be an undetermined value of the unknown and V_i the ξ residual for x_i; then

$$V_i = x_i - \xi.$$

Let $S(\xi)$ denote ΣV_i^2, an essentially positive quantity whatever the value of ξ may be. Then

$$S(\xi)=\Sigma x_i^2-2\xi\Sigma x_i+n\xi^2$$

$$=\Sigma x_i^2-2na\xi+n\xi^2 \tag{2}$$

by means of (1).

Similarly, if $S(a)$ denotes Σv_i^2, where v_i is the residual x_i-a, then on putting a for ξ in (2) we have

$$S(a)=\Sigma x_i^2-na^2. \tag{3}$$

From (2) and (3), $$S(\xi)-S(a)=n(\xi-a)^2;$$

hence, since the right-hand side is positive, $S(a)<S(\xi)$ or

$$\Sigma v_i^2<\Sigma V_i^2.$$

Thus, we have the *principle of least squares* which states that the sum of the squares of the residuals with respect to the arithmetic mean is the least possible.

Conversely, if we adopt, as the criterion for the best value of x, that value which makes the sum of the squares of the corresponding residuals a minimum, then we have to find the value of ξ such that $S(\xi)$ is a minimum. Now,

$$S(\xi)\equiv\Sigma V_i^2=\Sigma(x_i-\xi)^2,$$

from which $$\frac{\partial S}{\partial \xi}=-2\Sigma(x_i-\xi) \tag{4}$$

and $$\frac{\partial^2 S}{\partial \xi^2}=2n. \tag{5}$$

For a minimum, $\partial S/\partial\xi=0$ and $\partial^2 S/\partial\xi^2>0$; (5) shows that the second condition is satisfied; by means of (4) the first condition is

$$\Sigma(x_i-\xi)=0$$

or $$\Sigma x_i=n\xi;$$

accordingly, the value of ξ which makes $S(\xi)$ a minimum is the arithmetic mean. Thus, the principle of least squares leads to the result that the value of the unknown, consistent with this principle, is the arithmetic mean.

The postulate of the arithmetic mean and the principle of least squares are seen to be complementary, for one involves the other.

Note. In arithmetical work it is advisable to have one or more checks on the computations. Now, since $v_i = x_i - a$, then

$$\Sigma v_i = \Sigma x_i - na = 0. \tag{6}$$

One check is afforded by (6), which states that the algebraical sum of the residuals is zero.

2·12. Standard error

From 2·09 (1), $x_i - x = \epsilon_i \quad (i = 1, 2, \ldots, n)$,

or
$$x = x_i - \epsilon_i. \tag{1}$$

Also, if ϵ is the error of the arithmetic mean, a, then $a - x = \epsilon$ or

$$x = a - \epsilon. \tag{2}$$

Take the sum of the n equations (1); then

$$nx = \Sigma x_i - \Sigma \epsilon_i = na - \Sigma \epsilon_i,$$

from which, by (2),
$$n\epsilon = \epsilon_1 + \epsilon_2 + \ldots + \epsilon_n, \tag{3}$$

so that
$$n^2\epsilon^2 = \Sigma \epsilon_i^2 + \sum_{i=1}^{n} \sum_{j=1}^{n} \epsilon_i \epsilon_j \quad (j \neq i). \tag{4}$$

Set
$$\sum_{i=1}^{n} \sum_{j=1}^{n} \epsilon_i \epsilon_j = T \quad (j \neq i). \tag{5}$$

Then
$$n^2\epsilon^2 = \Sigma \epsilon_i^2 + T. \tag{6}$$

Let μ^2 denote the mean of the n quantities ϵ_i^2 so that

$$n\mu^2 = \Sigma \epsilon_i^2. \tag{7}$$

Then (6) becomes
$$\epsilon^2 = \frac{\mu^2}{n} + \frac{T}{n^2}. \tag{8}$$

The important positive quantity, μ, defined by (7) is analogous to the standard deviation defined in § 1·06 (iii). In the literature it is generally described as the *root-mean-square error*—a cumbrous expression. In the following pages μ will be described as the *standard error* (S.E.).

The equation (8) is an *exact* equation which gives the error, ϵ, of the arithmetic mean in terms of the errors of the individual observations. It is evident from the quadratic form of (8) that ϵ can be positive or negative, so that the true value of x in (2) is given by

$$x = a \pm |\epsilon|. \tag{9}$$

The incidence of the double sign will be discussed more fully in §2·14.

Consider now the double summation, T, given by (5), namely,

$$T=\sum_{i=1}^{n}\sum_{j=1}^{n}\epsilon_i\epsilon_j \quad (j\neq i).$$

The errors ϵ_i and ϵ_j, being accidental errors, may be positive or negative; consequently, T may be expected to be, numerically, much less than $\Sigma\epsilon_i^2$, the first term on the right-hand side of (6).

Let μ_a be the value of ϵ when T is neglected in (6); then (8) becomes $\mu_a^2=\mu^2/n$, or

$$\mu_a=\pm\frac{\mu}{\sqrt{n}}, \tag{10}$$

in which μ, as defined, is a positive quantity.

As we shall see more particularly in §2·14, the double sign in (10) reflects our ignorance as to which side of the arithmetic mean the true value lies.

Since T has been neglected, μ_a would ordinarily be regarded as an approximate value of the error of the arithmetic mean and, in a well-conducted series of observations, the degree of approximation would be expected to become closer with increasing values of n. In the ideal case when n is a very large number the formula (10) is to be regarded as giving the *theoretical* value of μ_a in terms of μ.

The numerical value of μ_a, that is, $|\mu_a|\equiv\mu/\sqrt{n}$, is the *standard error of the arithmetic mean* and, as we shall see later, it represents one index of precision, in use, of the arithmetic mean.

It is to be noted that in the present discussion the standard error, μ, defined by (7), is a *theoretical* quantity, since it is a function of the individual errors, ϵ_i, of observation which are themselves unknown. In the next section it will be shown how μ can be evaluated, usually with considerable accuracy, from the observed values, x_i, of the unknown; then the standard error, $|\mu_a|$, of the arithmetic mean can be found at once by means of (10).

As in (9) the solution for x is written with the double sign as

$$x=a\pm|\mu_a| \quad (\text{S.E.}),$$

in which (S.E.) is added to indicate that our discussion is in terms of standard error.

In the interests of simplicity of notation it will be sufficient to write the previous equation as

$$x=a\pm\mu_a \quad (\text{S.E.}), \tag{11}$$

in which μ_a *is to be interpreted as the standard error of the arithmetic mean*, namely, $|\mu_a|$ or $\mu/\sqrt{n}$.

2·13. The formula $\mu^2 = \dfrac{[vv]}{n-1}$

With the previous notation we have

$$x_i - x = \epsilon_i, \tag{1}$$

$$x_i - a = v_i, \tag{2}$$

$$a - x = \epsilon, \tag{3}$$

where ϵ is the error of the arithmetic mean.

From (1) and (2), $a - x = \epsilon_i - v_i$; hence, by (3),

$$v_i = \epsilon_i - \epsilon,$$

from which $\qquad \Sigma v_i^2 = \Sigma \epsilon_i^2 - 2\epsilon \Sigma \epsilon_i + n\epsilon^2 = n\mu^2 - n\epsilon^2$

by means of 2·12 (7) and 2·12 (3). Then, by 2·12 (8),

$$\Sigma v_i^2 = (n-1)\mu^2 - \frac{T}{n}. \tag{4}$$

This is an exact equation.

As before, it may be expected that T/n will be small compared with the first term on the right-hand side of (4), vanishing in the ideal case when n is very large. Neglecting T/n in (4) we then have the formula

$$\mu^2 = \frac{\Sigma v_i^2}{n-1}. \tag{5}$$

In the notation introduced by Gauss, Σv_i^2 is written as $[vv]$, signifying the sum of the products of v_i and v_i, that is, in this case, the sum of the squares of the residuals. We then write (5) as

$$\mu^2 = \frac{[vv]}{n-1}, \tag{6}$$

and 2·12 (10) becomes, with μ_a now denoting the standard error of the arithmetic mean as in 2·12 (11),

$$\mu_a \equiv \frac{\mu}{\sqrt{n}} = \sqrt{\left\{\frac{[vv]}{n(n-1)}\right\}}. \tag{7}$$

These are important formulae.

Because of their importance we summarize the main results and principles discussed in the previous pages.

(i) The standard error, μ, is a positive theoretical quantity given, in terms of the unknown errors of observation, by

$$\mu^2 = \frac{1}{n}\Sigma \epsilon_i^2.$$

If, for a series of observations, there is reason to believe that the observational errors are small, then μ will be small and the qualitative inference follows that the greater the degree of accuracy of a series of observations the smaller is the value of μ; thus the standard error, μ, is an index of precision relating to the magnitudes and distribution of the unknown errors.

(ii) The standard error, μ_a, of the arithmetic mean is similarly an index of precision; its theoretical value, that is, in terms of the unknown errors, is given by

$$\mu_a = \frac{\mu}{\sqrt{n}}.$$

(iii) In any series of observations the arithmetic mean and the residuals v_i can be evaluated by ordinary processes; consequently, the *computed* values of μ and μ_a can be obtained from the formulae (6) and (7) above.

(iv) The derivation of (6) as an *exact* formula depends on the vanishing of T/n on the right-hand side of (4). When n is large, T/n may be expected to be negligible compared with $[vv]$, in which event (6) and (7) can be regarded in practice as sufficiently exact formulae for determining the *theoretical* quantities μ and μ_a. If, however, n is small, say, $n=7$, the double summation can hardly be expected to vanish in general, and so (6) and (7) must be regarded as approximate formulae only. It follows that, in this case, the value of μ, given by $\sqrt{\left\{\frac{[vv]}{n-1}\right\}}$, is not an exact value of this important theoretical quantity, although, in a series of careful observations it is, as a rule, sufficiently precise for all practical purposes. Similar remarks apply to μ_a, given by $\mu/\sqrt{n}$.

(v) By (6), $$[vv] \equiv \Sigma v_i^2 = (n-1)\mu^2 < n\mu^2,$$

or, by 2·12 (7), $$\Sigma v_i^2 < \Sigma \epsilon_i^2. \tag{8}$$

Thus, on the whole, the residuals are less, numerically, than the true errors.

(vi) As in 2·12 (11), the solution is written as

$$x = a \pm \mu_a. \tag{9}$$

2·14. Illustrative example

We consider as a simple example the measurement, on an ordnance map, of the distance between two points A and B on a curved road. We can suppose that the map distance between A and B is found by laying a thread as accurately as possible along the curved line and translating the length of thread concerned into miles by means of the map scale. Further, we shall suppose that the operation has been performed n times by one or more persons, resulting in the values $d_1, d_2, \ldots, d_n$ (in miles) for the curvilinear road distance.

The distribution of the various values of d (to two places of decimals) in Table 6 has been taken in a special way, first, to make the calculations as simple as possible so as to focus attention on methods of procedure and the principles involved; secondly, the distribution has been taken in accordance with the general experience that small errors are more numerous than large errors; thirdly, the statistics are intended to illustrate the relevance of the double sign in the formula

$$x = a \pm \mu_a.$$

The equation associated with the measure d_i is

$$d_i - d = e_i \quad (i = 1, 2, \ldots, n), \tag{1}$$

where d is the *true* value of the distance and e_i is the accidental error of measurement converted into miles. In the table, $n = 13$.

Table 6

(1)	(2)	(3) ϵ_i		(4)	(5)	(6) $\epsilon_i U_i$		(7)	(8)
d_i	x_i	+	−	ϵ_i^2	U_i	+	−	v_i	v_i^2
2·38	−6		5	25	18		90	−6	36
2·40	−4		3	9	16		48	−4	16
2·40	−4		3	9	16		48	−4	16
2·42	−2		1	1	14		14	−2	4
2·42	−2		1	1	14		14	−2	4
2·44	0	1		1	12	12		0	0
2·44	0	1		1	12	12		0	0
2·44	0	1		1	12	12		0	0
2·46	+2	3		9	10	30		+2	4
2·46	+2	3		9	10	30		+2	4
2·48	+4	5		25	8	40		+4	16
2·48	+4	5		25	8	40		+4	16
2·50	+6	7		49	6	42		+6	36
	0	**+13**		**165**	**156**	**218**	**214**	**0**	**152**

Summary: $n = 13$; $\Sigma x_i = 0$; $\Sigma \epsilon_i = 13$; $\Sigma \epsilon_i^2 = 165$; $\Sigma U_i = 156$; $\Sigma \epsilon_i U_i = 4$; $\Sigma v_i = 0$; $\Sigma v_i^2 \equiv [vv] = 152$.

In numerical work it is always advisable to reduce a basic equation, such as (1), to as simple a form as possible by cutting out unnecessary numerical components and by avoiding decimals. In the present example it is easily seen from the first column of the table that, owing to the intentional symmetry of the values of d_i, the arithmetic mean, a, is given by

$$a = 2{\cdot}44. \tag{2}$$

It is then advantageous to write

$$d_i = a + cx_i = 2{\cdot}44 + cx_i, \tag{3}$$

in which $c = 0{\cdot}01$ mile, and, consequently, the values of x_i are positive or negative integers; these values are shown in the second column. In a real problem in which symmetry is unlikely to be a feature of the distribution we should choose, instead of a, a convenient value, a_0, in the neighbourhood of the estimated mean.

Let x be the true value corresponding to the true distance, d, so that

$$d = a + cx = 2{\cdot}44 + cx. \tag{4}$$

Let ϵ_i denote the error of x_i so that

$$\epsilon_i = x_i - x; \tag{5}$$

then, from (1), (3) and (4), $\quad e_i = c\epsilon_i;$

the errors of d_i are thus related simply to the errors of x_i.

The problem of dealing with the distances d_i is now transformed into the problem of dealing with the numbers x_i.

If e denotes the error of a, then

$$e = a - d. \tag{6}$$

Also, from the second column of the table,

$$\bar{x} \equiv \frac{1}{n}\Sigma x_i = 0; \tag{7}$$

hence, if ϵ denotes the error of $\bar{x}$, then

$$\epsilon \equiv \bar{x} - x = -x. \tag{8}$$

To illustrate general principles *it will now be assumed that the true value, d, of the distance is* 2·43 *miles*. Then, from (4),

$$x = -1;$$

from (5) and (8), $\quad \epsilon_i = x_i + 1 \quad$ and $\quad \epsilon = +1;$

also, from (2) and (6), $\quad e = +1{\cdot}00.$

The values of ϵ_i are shown in the third column and their squares in the fourth.

(i) In this subsection we proceed to calculate the double summation, T, given by

$$T = \sum_{i=1}^{n} \epsilon_i \sum_{j=1}^{n} \epsilon_j \quad (j \neq i).$$

Let $$s = \Sigma\epsilon_i = 13 \tag{9}$$

and $$U_i = s - \epsilon_i = 13 - \epsilon_i. \tag{10}$$

Then $$T = \Sigma\epsilon_i(s - \epsilon_i) = \Sigma\epsilon_i U_i.$$

The values of U_i and $\epsilon_i U_i$ are in columns (5) and (6); we then find that $T = 218 - 214 = 4$. Thus, T is small compared with the value, 165, of $\Sigma\epsilon_i^2$, in accordance with the expectation expressed in § 2·12.

(ii) The residuals $v_i \equiv x_i - \bar{x} = x_i$ (since $\bar{x} = 0$) and their squares are given in the last two columns of the table. It will be observed that $\Sigma v_i^2 < \Sigma\epsilon_i^2$, in accordance with the formula 2·13 (8).

(iii) It is of utmost importance to apply checks for the various steps in the computations; the methods of doing so depend on the particular ways in which the calculations are carried out. Despite the simplicity of the numerical work in Table 6 we consider two checks.

First, by means of (10) and (9),

$$\Sigma U_i \equiv ns - \Sigma\epsilon_i = 169 - 13 = 156,$$

as shown at the bottom of the fifth column. Secondly, by 2·11 (6), $\Sigma v_i = 0$; the summary shows that this result is verified.

(iv) We now test the accuracy of the formula 2·13 (6), denoting the value of μ, calculated according to this formula, by μ_c; then

$$\mu_c^2 \equiv \frac{[vv]}{n-1} = \frac{152}{12},$$

from which $$\mu_c = 3{\cdot}57.$$

The *true* value, μ, is given by $n\mu^2 = \Sigma\epsilon_i^2$ or by $13\mu^2 = 165$, from which

$$\mu = 3{\cdot}56.$$

Thus, the values of μ_c and μ are practically identical. The almost exact accordance is, of course, the result of the artificial character of the data in the first column of Table 6; in a real series of careful measurements it would, however, be expected that the value of μ_c would not differ significantly from the true value, μ.

(v) The value of μ_a, calculated from the residuals, follows from the formula $\mu_a = \mu_c/\sqrt{n}$; hence

$$\mu_a = 3{\cdot}57/\sqrt{13} = 0{\cdot}987.$$

As previously stated the error, ϵ, of the arithmetic mean of the x's, that is, of $\bar{x}$ which is zero, by (7), is

$$\epsilon = +1.$$

Thus, the values of μ_a and ϵ are practically identical, showing that, numerically, the value of μ_a represents the true error of $\bar{x}$ with a very high degree of accuracy.

The unit in which μ_c, μ, ϵ and μ_a are expressed is 0·01; hence, in miles, and to three places of decimals,

$$a = 2{\cdot}440, \quad \mu_c = \mu = 0{\cdot}036,$$

$$e\,(\text{error of } a) = +0{\cdot}010, \quad \mu_a = 0{\cdot}010.$$

From (6), $$d = a - e,$$

and, since $e = \mu_a = 0{\cdot}010$, we have the identity

$$d \equiv a - \mu_a, \tag{11}$$

where d ($\equiv 2{\cdot}43$) is the *assumed* true distance and $a = 2{\cdot}44$.

(vi) Consider now the same measured values, d_i, in Table 6 *but assume now that the true value of the distance, denoted by* d_1, *is* 2·45.

The arithmetic mean, a, is 2·44 as before, and, by (3), the values of x_i in column (2) are unaltered. The error of a, now denoted by e_1, is given by

$$e_1 \equiv a - d_1 = -0{\cdot}010. \tag{12}$$

By (4), the true value, x, is now $+1$, and, from (5), the error ϵ_i is given by

$$\epsilon_i = x_i - 1.$$

Thus the errors, ϵ_i, are -7, -6, ..., which are the errors in column (3) with their signs changed but reading upwards from the bottom of the column. It follows that $\Sigma\epsilon_i^2$ is unchanged in value and that the value of μ is unaltered. Further, since the residuals are independent of any knowledge as to the true value of the unknown, being dependent primarily on the arithmetic mean, the columns (7) and (8) remain unchanged; hence, the calculated value of μ_a is the same as before, namely,

$$\mu_a = 0{\cdot}010,$$

so that, since $e_1 = -0{\cdot}010$, then

$$\mu_a = -e_1.$$

Now, by (12), $d_1 = a - e_1$ and we now have the identity

$$d_1 \equiv a + \mu_a. \tag{13}$$

(vii) If we now suppose that the true value of the distance is *either* 2·43 *or* 2·45, the formulae (11) and (13) combine to give the true distance as

$$d = a \pm \mu_a, \tag{14}$$

in which μ_a has been identified, to the degree of accuracy indicated in our problem, with the true error (numerical) of the arithmetic mean.

The double sign in (14) merely reflects our ignorance as to which of the two values, 2·43 and 2·45, represents the unique value of the unknown. Further, the value of μ_a, being calculated from the residuals, is independent of any knowledge as to the true value of the unknown.

(viii) In the general problem involving measures X_i, when of course the true value, X, is not known, the *best* value to be attributed to the unknown is, according to the principle of least squares, the arithmetic mean, a, of the measures. The precision, μ_a, of the arithmetic mean is calculated from the formula

$$\mu_a = \sqrt{\left\{\frac{[vv]}{n(n-1)}\right\}},$$

and the solution is then written as

$$X = a \pm \mu_a \quad \text{(S.E.)},$$

in which μ_a *may be only an approximate value of the error* (numerical) of the arithmetic mean.

2·15. Observations in which a constant systematic error is present

We denote by k a constant systematic error which enters into all the measures x_i of a true quantity x; the arithmetic mean is a.

Let ϵ_i denote the *accidental* error of the measure x_i; then

$$x_i - x = \epsilon_i + k \quad (i = 1, 2, \ldots, n),$$

from which, by summation,

$$\Sigma x_i - nx \equiv n(a - x) = \Sigma\epsilon_i + nk. \tag{1}$$

Let ϵ denote the resultant error of the arithmetic mean, which will be partly accidental and partly systematic; then

$$a - x = \epsilon.$$

Hence (1) becomes $\qquad n\epsilon = \Sigma\epsilon_i + nk;$
on squaring we have

$$\begin{aligned} n^2\epsilon^2 &= (\Sigma\epsilon_i)^2 + n^2k^2 + 2nk\Sigma\epsilon_i \\ &= \Sigma\epsilon_i^2 + T + n^2k^2 + 2nks, \end{aligned} \tag{2}$$

where $\qquad T = \sum_{i=1}^{n} \sum_{j=1}^{n} \epsilon_i \epsilon_j \quad (j \neq i),$

and $\qquad s = \Sigma\epsilon_i.$

Now, as we have seen earlier, T may be expected to be small since the accidental errors, ϵ_i, may be positive or negative, and the same remark applies equally to s. If μ_a is the value of ϵ when T and s are neglected and if μ is the standard error associated with the accidental errors and given by $n\mu^2 = \Sigma\epsilon_i^2$, then (2) becomes

$$\mu_a^2 = \frac{\mu^2}{n} + k^2. \tag{3}$$

Since each x_i contains the constant error k, it follows that the arithmetic mean will also contain this constant error; hence the residuals v_i depend only on the accidental errors. Thus, the standard error, μ, of the accidental errors is given by the formula

$$\mu^2 = \frac{[vv]}{n-1},$$

whether a constant systematic error is present or not. Hence, from (3),

$$\mu_a^2 = \frac{[vv]}{n(n-1)} + k^2.$$

However diligently we increase the number of observations, thereby tending to reduce the effect of the accidental errors, the value of μ_a^2 will never be less than k^2.

To obtain a reliable value of the unknown when a systematic error is suspected to be present, it is necessary, first, to make a subsidiary series of observations or experiments designed to evaluate the constant error k and, second, to remove k from each of the measured quantities x_i.

2·16. The standard error of a linear function of two independent variables

Consider the function ϕ given by

$$\phi = ax + by, \tag{1}$$

where x and y are independent variables, and a and b are constants. Suppose that A makes a series of n observations of the quantity x and B makes a similar series of n observations of y, a particular observation x_i being associated with y_i. The corresponding value of ϕ, namely, ϕ_i, is given by

$$\phi_i = ax_i + by_i.$$

If $\bar{\phi}$ is the mean value of the n values ϕ_i, then

$$\bar{\phi} = a\bar{x} + b\bar{y}.$$

Let ξ_i and η_i be the accidental errors in x_i and y_i, and ϵ_i the corresponding error in ϕ_i. Then

$$\epsilon_i = a\xi_i + b\eta_i,$$

from which

$$\Sigma\epsilon_i^2 = a^2\Sigma\xi_i^2 + b^2\Sigma\eta_i^2 + 2ab\Sigma\xi_i\eta_i. \tag{2}$$

At this point we introduce the important stipulation that the errors made by A are entirely independent of the errors made by B; in other words, we assume that, for a given ξ_i, the corresponding η_i can be positive or negative and that the magnitude of η_i is independent of the magnitude of ξ_i. It follows that $\Sigma\xi_i\eta_i$ will consist of positive and negative quantities and, in the ideal case when n is large, that the summation may be expected to be small or negligible compared with $\Sigma\xi_i^2$ or $\Sigma\eta_i^2$.

Let μ_ϕ, μ_x and μ_y denote the standard errors of ϕ, x and y; then

$$n\mu_\phi^2 = \Sigma\epsilon_i^2, \quad n\mu_x^2 = \Sigma\xi_i^2 \quad \text{and} \quad n\mu_y^2 = \Sigma\eta_i^2.$$

Then, neglecting the final term in (2), we have

$$\mu_\phi^2 = a^2\mu_x^2 + b^2\mu_y^2. \tag{3}$$

This result can obviously be generalized for a linear function of any number of independent variables; in particular, if $\phi = ax + by + cz$, then

$$\mu_\phi^2 = a^2\mu_x^2 + b^2\mu_y^2 + c^2\mu_z^2.$$

We shall meet again the subject-matter of this section when, in § 4·14, we base our arguments on the principles of probability.

Corollary. If ϕ denotes the mean of x and y, then

$$\mu_\phi = \tfrac{1}{2}\sqrt{(\mu_x^2 + \mu_y^2)}.$$

2·17. Hypothesis of elementary errors

It will be assumed that the errors affecting the measurement of a particular characteristic are accidental errors only. In practice, generally, there is some undefined limit to the magnitude of such errors; thus, in measuring the length of a curved line, as in § 2·14, it could be confidently anticipated that no reasonably expert measurer would or could be so careless as to make an error of half a mile, say, in any one of his measures; in other words, the frequency of such an error is negligible. Further, we should expect the frequency of small errors to be greater than the frequency of larger errors. If a law of errors exists, it must take cognizance of what is a matter of experience.

In considering the problem of deriving a law of errors based on the simple concepts of this chapter, it is necessary to introduce an *ad hoc* assumption or hypothesis which is reasonable in the light of experience. The hypothesis adopted in the present connexion is that a particular error in an observation is made up of a large number of small but discrete *elementary errors*, all of the same magnitude, some positive and some negative. In the case of determining the right ascension of a star, for example, there is a large number of individual accidental errors arising from all sorts of sources and contributing in different ways to produce the final but unknown error associated with the observed value of the right ascension; the hypothesis thus states that this final error is a *particular* combination of positive and negative elementary errors.

The hypothesis was introduced originally by Hagen in 1837 as the basis on which the normal law of errors, as it is called, was derived.

2·18. The normal law of errors

We denote the magnitude of an elementary error by ϵ and we assume that all accidental errors of observation are derivable from n elementary errors by taking various combinations of signs. Thus, a particular accidental error, ξ, of observation consists of a certain number, m, of negative elementary errors and the number, $n-m$, of positive elementary errors; then

$$\xi = m(-\epsilon) + (n-m)\,\epsilon = (n-2m)\,\epsilon. \tag{1}$$

The number of ways in which ξ can be produced, denoted by $F(\xi)$, is the number of ways of selecting m elementary errors out of the total n; accordingly,

$$F(\xi) = \frac{n!}{m!(n-m)!}. \tag{2}$$

In the circumstances stated, $F(\xi)$ is the frequency of the error ξ.

Similarly, an error $(\xi+2\epsilon) \equiv (n-2m+2)\,\epsilon$ will be produced by combining $(m-1)$ negative elementary errors with $(n-m+1)$ positive elementary errors; hence

$$F(\xi+2\epsilon) = \frac{n!}{(m-1)!\,(n-m+1)!}. \tag{3}$$

From (2) and (3),
$$\frac{F(\xi+2\epsilon)}{F(\xi)} = \frac{m}{n-m+1},$$

from which

$$\frac{F(\xi+2\epsilon)-F(\xi)}{F(\xi+2\epsilon)+F(\xi)} = -\frac{n-2m+1}{n+1} = -\frac{(\xi+\epsilon)}{(n+1)\,\epsilon}, \tag{4}$$

by means of (1).

If the number of elementary errors is very large and $F(\xi)$ is assumed to be a continuous function then, to $O(\epsilon)$,

$$F(\xi+2\epsilon) \equiv F(\overline{\xi+\epsilon}+\epsilon) = F(\xi+\epsilon) + \epsilon F'(\xi+\epsilon)$$

and
$$F(\xi) \equiv F(\overline{\xi+\epsilon}-\epsilon) = F(\xi+\epsilon) - \epsilon F'(\xi+\epsilon).$$

Then, (4) becomes, to $O(\epsilon)$,

$$\frac{\epsilon F'(\xi+\epsilon)}{F(\xi+\epsilon)} = -\frac{\xi+\epsilon}{(n+1)\,\epsilon},$$

or, if x denotes the error $\xi+\epsilon$,

$$\frac{F'(x)}{F(x)} = -\frac{x}{(n+1)\epsilon^2}.$$

It is now assumed that the number of elementary errors and their magnitudes are such that $(n+1)\epsilon^2 = 1/2h^2$, where h is a constant. Then

$$\frac{F'(x)}{F(x)} \equiv \frac{d}{dx}\{\log F(x)\} = -2h^2 x,$$

from which, on integration,

$$F(x) = A\, e^{-h^2x^2}, \tag{5}$$

where A is a constant of integration.

The frequency of errors between x and $x+dx$ is then

$$F(x)\,dx \equiv A\, e^{-h^2x^2}\,dx, \tag{6}$$

and if all errors between $-p$ and $+p$ are possible then the total frequency of errors, $S(p)$, is given by

$$S(p) = A\int_{-p}^{p} e^{-h^2x^2}\,dx. \tag{7}$$

If generous limits are allowed to the maximum errors $\pm p$, the total frequency $S(p)$ differs little from the total frequency, S, when the limits in (7) are taken to be $-\infty$ and $+\infty$, for, as in § 2·17, we regard the frequency of very large errors to be extremely small. Then, with sufficient accuracy,

$$S = A\int_{-\infty}^{\infty} e^{-h^2x^2}\,dx.$$

From 1·14 (20), the value of the integral is $\sqrt{\pi}/h$; hence

$$S = A\sqrt{\pi}/h. \tag{8}$$

The *relative frequency* of errors between x and $x+dx$ is denoted by $f(x)\,dx$; then, by (6) and (8), $f(x) = F(x)/S$, so that

$$f(x) = \frac{h}{\sqrt{\pi}}\, e^{-h^2x^2}. \tag{9}$$

The function $f(x)$, given by (9), is the *normal function*, or *Gaussian function*, associated with the law of errors.

The normal law of errors may then be stated in the form: the relative frequency of errors between x and $x+dx$ is

$$f(x)\,dx \equiv \frac{h}{\sqrt{\pi}}\, e^{-h^2x^2}\,dx. \tag{10}$$

The constant, h, appearing in the normal function is called the *modulus of precision*.

If the number, N, of observations is large, the frequency of errors between x and $x+dx$ is

$$N\frac{h}{\sqrt{\pi}}e^{-h^2x^2}\,dx. \tag{11}$$

2·19. The modulus of precision

From 2·18 (10) it is at once deduced that the relative frequency of all errors within the range $-x$ to $+x$ is

$$\frac{h}{\sqrt{\pi}}\int_{-x}^{x}e^{-h^2x^2}\,dx \quad\text{or}\quad \frac{2h}{\sqrt{\pi}}\int_{0}^{x}e^{-h^2x^2}\,dx. \tag{1}$$

Write $t=hx$; the limits for t are 0 and hx in the second form of (1). The relative frequency, defined by (1), is then

$$\frac{2}{\sqrt{\pi}}\int_0^{hx} e\ {t^2}\,dl.$$

In § 1·15 (10) we introduced the *error function*, erf t, defined by

$$\text{erf}\,(t)=\frac{2}{\sqrt{\pi}}\int_0^t e^{-t^2}\,dt;$$

it may be remarked that its name arose in the present connexion.

Hence, the relative frequency of all errors in the range $-x$ to x is

$$N\,\text{erf}\,(hx). \tag{2}$$

Also, the number of errors between 0 and x is $\frac{1}{2}N\,\text{erf}\,(hx)$.

Consider two series of observations of the same quantity, with h and h_1 as the moduli of precision, and suppose that $h>h_1$. Then, for any *prescribed range* $-x$ to $+x$ for the errors, the relative frequencies are $\text{erf}\,(hx)$ and $\text{erf}\,(h_1x)$. Now, the integrand e^{-t^2} in (2) or (3) is positive and hence, since $h>h_1$, $\text{erf}\,(hx)>\text{erf}\,(h_1x)$, that is, the relative frequency in the first case (h) is greater than the relative frequency in the second case (h_1). In other words, the errors are distributed more closely about $x=0$ in the first case than in the second case. This implies that the series of observations in the first case is more accurate than the series of observations in the second case. Thus, we have the general principle that the more accurate a series of observations is, the greater is the value of h; hence, h is an index, or modulus, of precision associated with a series of observations.

The two graphs of the normal function are shown in Fig. 9, in which $OA=h/\sqrt{\pi}$ and $OA_1=h_1/\sqrt{\pi}$. Since we are dealing with relative frequency, the area under each curve is unity.

It is to be noted that the function relating to the normal law of errors is the same as that for the normal distribution function discussed in §§ 1·13, 1·15 and illustrated in Fig. 6.

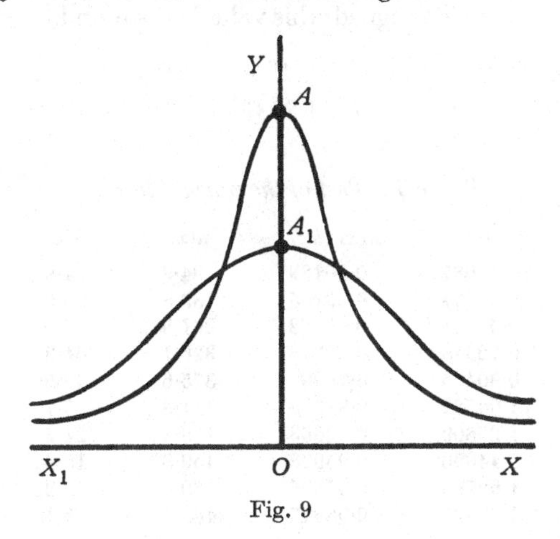

Fig. 9

2·20. Test of the normal law of errors

The theoretical derivation of a law of errors carries with it an obligation to submit the law to the test offered by a series of observations which have been made with the greatest care and from which it is assumed that the effects of instrumental, personal and systematic errors have been removed; in this category are, for example, a long series of astronomical observations and the measures of physical constants made in the laboratory. A condition for the reliability of a test of the normal law is that the observations or measures should be numerous.

We consider the results of 470 observations† by Bradley of the right ascensions of *Sirius* and *Altair*, expressed in seconds of arc. As the number of observations is very considerable the residuals from the arithmetic mean for each star may be regarded as the errors of the individual observations. The accuracy of instruments in the time of Bradley was very much inferior to that in modern days and, accordingly, comparatively large errors would be expected in the observations; perhaps, for this reason, the statistics to be considered afford a thorough test of the normal law.

In the last column of Table 7 are to be found the observed number of errors, taken without regard to sign, between $0''\cdot 0$ and $0''\cdot 1$, $0''\cdot 1$

† W. Chauvenet, *Spherical and Practical Astronomy*, 5th ed. **2** (Philadelphia, 1891), p. 489.

and $0''\cdot 2$, and so on; these numbers, denoted by $\Delta_0(x)$, will be compared with the theoretical numbers derived on the assumption that the errors follow the normal law to which a particular value of the modulus of precision is assigned; this value† was found to be given by

$$\frac{1}{h}=\frac{1''}{1\cdot 8087}. \tag{1}$$

Table 7. *Test of the normal law*

x	hx	erf (hx)	$n(x)$	$\Delta(x)$	$\Delta_0(x)$
$0''\cdot 1$	0·18087	0·20189	94·9	94·9	94
$0''\cdot 2$	0·36174	0·39105	183·8	88·9	88
$0''\cdot 3$	0·54261	0·55712	261·8	78·0	78
$0''\cdot 4$	0·72348	0·69376	326·1	64·3	58
$0''\cdot 5$	0·90435	0·79908	375·6	49·5	51
$0''\cdot 6$	1·08522	0·87515	411·3	35·7	36
$0''\cdot 7$	1·26609	0·92663	435·5	24·2	26
$0''\cdot 8$	1·44696	0·95928	450·8	15·3	14
$0''\cdot 9$	1·62783	0·97866	460·0	9·2	10
$1''\cdot 0$	1·8087	0·98947	465·1	5·1	7
$>1''\cdot 0$	—	—	—	4·9	8

If N is the total number of observations—in this case, $N=470$—the frequency, $n(x)$, of errors between $-x$ and $+x$ is given, from 2·19 (2), by

$$n(x)=N\,\mathrm{erf}\,(hx).$$

If $x=0''\cdot 1$ then, by (1),

$$hx=\frac{0''\cdot 1}{1/h}=0\cdot 18087; \tag{2}$$

hence from the table of the error function (Appendix 1) we have

$$n(0''\cdot 1)\equiv N\,\mathrm{erf}\,(0\cdot 18087)=0\cdot 20189N=94\cdot 9.$$

It is to be noticed that hx in (2) is a number. Similarly, if $x=0''\cdot 2$, then $hx=0\cdot 36174$ and

$$n(0''\cdot 2)=N\,\mathrm{erf}\,(0\cdot 36174)=183\cdot 8.$$

The other values of $n(x)$ are found in the same way.

Let $\Delta(x)$ denote the number of errors between $x-0''\cdot 1$ and x; thus

$$\Delta(x)=n(x)-n(x-0''\cdot 1).$$

† What is called the *probable error*, r, was found by Bessel to be given by $r=0''\cdot 2637$; in chapter 4 we shall see that r and h are connected by the formula $r=0\cdot 4769(1/h)$, from which (1) is derived.

The values of $\Delta(x)$ are found in the fifth column; these values give the *theoretical* numbers of errors between $0''\cdot0$ and $0''\cdot1$, between $0''\cdot1$ and $0''\cdot2$, and so on. As previously mentioned the last column contains the *observed* numbers, $\Delta_0(x)$, in these ranges.

The agreement between the implications of the normal law and the observational errors as shown in the last two columns is remarkably close on the whole; for errors up to $1''\cdot0$ the only noteworthy divergence is for $x=0''\cdot4$. It is to be noted, however, that the number of observed errors exceeding $1''\cdot0$ numerically is somewhat greater than the theoretical number; this is in fairly general accordance with most investigations of a similar nature, suggesting that the law of errors tends slightly to under-estimate the proportion of large errors. In § 8·01 we shall have occasion to refer to this point again.

CHAPTER 3

PROBABILITY AND THE NORMAL LAW

3·01. Probability

In this section and the next we consider the basic concepts of probability by means of which the normal law will be derived.

In tossing a coin there are two possibilities as to which side will fall uppermost, 'heads' or 'tails', and, on the assumption that the coin is perfectly symmetrical, there is no presumptive reason why the coin should fall 'heads' rather than 'tails'. If one should contemplate tossing the coin, say, a hundred times, the *a priori* expectation is that it will fall 'heads' fifty times and 'tails' fifty times; here 'reason' is independent of the necessity for performing the operation of tossing the coin a hundred times. The *Laplacian definition of probability* is based on these concepts and is stated in the form: if an event (for example, the fall of 'heads') is expected to occur m times in n trials, the probability, p, of the event occurring is defined to be given by $p=m/n$; the probability, q, of the event *not* occurring is given by $q=(n-m)/n$; thus, $p+q=1$.

The probability of a coin falling 'heads' is then 1/2, and the probability of its falling 'tails' is also 1/2.

In the case of a cubical die the probability of throwing a particular side, say 'two', is 1/6, since the total number of possibilities is 6 (the number of faces of the cube).

An alternative definition of probability is based on experiment. Suppose that a large number, N, of trials are made, as in tossing a coin or throwing a cubical die, and that a particular event, such as falling 'heads' or falling 'two', occurs pN times, then the probability of the event is said to be p. The frequency of the event is pN. In the past several extensive trials of tossing a coin have been made. For example, Buffon, tossed a coin 4040 times and it fells 'heads' 2048 times. The probability of falling 'heads' according to the *empirical definition* is, in this case, 2048/4040 or 0·507, which is close to the *a priori* estimate in the Laplacian definition that the probability should be exactly 0·5.

It is to be noted that if an event is *certain* to occur, the corresponding probability, p, is 1.

3·02. Compound probability

We first consider a simple example. Suppose X throws a coin and Y, simultaneously or immediately after, throws a die. What is the probability that X throws 'heads' (event A) and Y throws 'two' (event B)? The events A and B are independent and we regard the occurrence of the two events together as a compound event C.

Generalizing, suppose that the probability of the occurrence of event A is p_1 ($\equiv m_1/n_1$) and that the probability of the occurrence of event B, *independent of* A, is p_2 ($\equiv m_2/n_2$); it is assumed, as in the Laplacian definition, that the m's and n's are integers. In $(n_1 n_2)$ trials the number of times the event A is expected to occur, is $p_1(n_1 n_2)$ or $(m_1 n_2)$, and of these the number of times the event B is expected to occur is $p_2(m_1 n_2)$ or $(m_1 m_2)$. Thus, in $(n_1 n_2)$ trials the events A and B occur together, that is, the compound event C occurs, $(m_1 m_2)$ times. Hence, the probability, p, of the occurrence of the compound event C is $m_1 m_2/n_1 n_2$; so that

$$p = p_1 p_2.$$

This result can evidently be generalized when n independent events are concerned. The probability, P, that these events occur together is given by

$$P = p_1 p_2 \ldots p_n.$$

3·03. Gauss's derivation of the normal law of errors

Gauss started with the postulate of the arithmetic mean, stated now in terms of probability as follows.

The arithmetic mean of the independent observations x_i ($i = 1, 2, \ldots, n$) of a quantity x—made under comparable conditions and deemed to be equally trustworthy—is the *most probable value* of the unknown, x.

The error of the observation x_i is ϵ_i, given by

$$x_i - x = \epsilon_i,$$

where x is the true value of the unknown.

If c represents the accuracy of measurement of the instrument employed, the probability of an error occurring between $\epsilon_i - \frac{1}{2}c$ and $\epsilon_i + \frac{1}{2}c$ will be proportional to c and will also be a function of ϵ_i, or of $x_i - x$; thus, the probability concerned is written as $cf(x_i - x)$. By the principle of compound probability (§ 3·02), the probability that all the independent errors occur is

$$P \equiv c^n f(x_1 - x) f(x_2 - x) \ldots f(x_n - x). \tag{1}$$

We take as the *most probable* value of the unknown that value which makes P a maximum; the conditions are (i) $dP/dx = 0$ and (ii) $d^2P/dx^2 < 0$.

By logarithmic differentiation of (1), the first condition is

$$\Sigma\left(\frac{f'(x_i - x)}{f(x_i - x)}\right) = 0,$$

or, if
$$F(x_i - x) \equiv \frac{f'(x_i - x)}{f(x_i - x)}, \tag{2}$$

then
$$\Sigma F(x_i - x) = 0. \tag{3}$$

Introduce now the postulate of the arithmetic mean, a; then the value of x in (3) is to be identified with a so that, if v_i is the residual $(x_i - a)$, (3) becomes

$$V \equiv F(v_1) + F(v_2) + \ldots + F(v_n) = 0. \tag{4}$$

Now, $\Sigma x_i = na$ or $\Sigma(x_i - a) = 0$ or $\Sigma v_i = 0$;

the last formula can be written as

$$v_n = -(v_1 + v_2 + \ldots + v_{n-1}), \tag{5}$$

so that, in (4), V becomes a function of $n-1$ independent v's, namely, $v_1, v_2, \ldots, v_{n-1}$. Hence $\partial V / \partial v_1 = 0$, that is,

$$\frac{\partial F(v_1)}{\partial v_1} + \frac{\partial F(v_n)}{\partial v_n} \cdot \frac{\partial v_n}{\partial v_1} = 0,$$

or, by (5),
$$\frac{\partial F(v_1)}{\partial v_1} = \frac{\partial F(v_n)}{\partial v_n}.$$

This last equation is evidently true for any pair of v's and, consequently, each differential coefficient must be equal to a constant, say, k. If v denotes any residual, then

$$\frac{\partial F(v)}{\partial v} = k,$$

from which
$$F(v) = kv + K.$$

Hence
$$V = k\Sigma v_i + nK.$$

But $V = 0$ and $\Sigma v_i = 0$; hence $K = 0$.

Now, from (2),
$$F(v) = \frac{f'(v)}{f(v)},$$

and we now have
$$\frac{f'(v)}{f(v)} = kv,$$

from which
$$f(v) = C\, e^{\frac{1}{2}kv^2},$$

where C is a constant of integration. The last result gives the *form* of the function f and so, returning to (1), we have

$$P = (cC)^n\, e^{\frac{1}{2}k\Sigma(x_i - x)^2}.$$

If the second condition for a maximum is to be fulfilled, that is, $d^2P/dx^2<0$, then it is easily seen that k must be negative. Put $k=-2h^2$; then

$$f(x_i-x)=C\,e^{-h^2(x_i-x)^2}. \tag{6}$$

In terms of the error, ϵ_i, or ϵ in short, the function $f(\epsilon)$ is given by

$$f(\epsilon)=C\,e^{-h^2\epsilon^2}. \tag{7}$$

We shall see later that $C=h/\sqrt{\pi}$; then $f(\epsilon)$ is the function associated with the normal law of errors.

3·04. The postulate of the arithmetic mean

In the preceding section the postulate that the arithmetic mean is the most probable value of the unknown leads to the form of the function f. The converse follows, namely, that if the form of the function f for discrete observations is given by 3·03 (6), then the most probable value of the unknown is the arithmetic mean. From 3·03 (1) and 3·03 (6),

$$P=(cC)^n\,e^{-h^2\Sigma(x_i-x)^2}.$$

Now, P is a maximum if $S\equiv\Sigma(x_i-x)^2$ is a minimum, that is, if (i) $\partial S/\partial x=0$ and (ii) $\partial^2S/\partial x^2>0$. The first condition is $\Sigma(x_i-x)=0$, and the second condition is satisfied since $\partial^2S/\partial x^2=2n$. The first condition is equivalent to

$$\Sigma x_i=nx,$$

or, stated otherwise, the most probable value of the unknown is the arithmetic mean.

As in § 2·11, the postulate of the arithmetic mean and the normal law are complementary, for one involves the other.

3·05. Continuous probability

It is to be remarked that the form of the function f in 3·03 (7) has been derived in connexion with a finite number of discrete observations and, further, that the constant of integration, C, is so far undetermined, the only property of C which we can assert at present being that C is positive for, since f is positive, we cannot have 'negative probability'.

If the number of observations is very great we can pass to the conception of *continuous probability* by defining the probability that an error lies between ϵ and $\epsilon+d\epsilon$ as $f(\epsilon)\,d\epsilon$, where f is given by 3·03 (7). Now, the probability of making *all* errors in the range $-\infty<\epsilon<\infty$ is unity, which is simply another way of asserting that it is *certain* that the errors lie within the above range. Hence

$$\int_{-\infty}^{\infty} f(\epsilon)\, d\epsilon = 1$$

or

$$C\int_{-\infty}^{\infty} e^{-h^2\epsilon^2} d\epsilon = 1.$$

From 1·14 (20), the value of the integral is $\sqrt{\pi}/h$; hence $C = h/\sqrt{\pi}$, and

$$f(\epsilon) = \frac{h}{\sqrt{\pi}} e^{-h^2\epsilon^2}. \tag{1}$$

The normal law of errors is then stated in the form: in a large number of observations the probability that an error lies between ϵ and $\epsilon + d\epsilon$ is

$$\frac{h}{\sqrt{\pi}} e^{-h^2\epsilon^2} d\epsilon. \tag{2}$$

We sometimes refer to the law as the *Gaussian law of errors*.

By means of the empirical definition of probability (§ 3·01) we can pass to the probability of the *frequency* of errors within the range ϵ to $e + de$. Thus, if the number, N, of observations is very large, the probable frequency of errors between ϵ and $\epsilon + d\epsilon$, which we denote by $n(e)\, de$, is given by

$$n(\epsilon)\, d\epsilon = N \frac{h}{\sqrt{\pi}} e^{-h^2\epsilon^2} d\epsilon. \tag{3}$$

3·06. Herschel's derivation of the normal law

Sir John Herschel considered the nature of the errors in dropping shot from a height on to a horizontal board with a particular point, O, indicated as that to be aimed at.

In Fig. 10 let X_1OX, Y_1OY be any arbitrary system of rectangular axes in the plane of the board. Let (x, y) be the coordinates of the point, P, of impact on the board of a particular shot; then x is the error parallel to OX and y is the error parallel to OY.

We now make the assumptions (i) that, inasmuch as the axes are arbitrary, the law of errors will be expressed in terms of the same function, f, for y as for x, and (ii) that the law of errors for x is independent of whatever values the error y may have, and vice versa.

The probability that an error parallel to OX lies between x and $x + dx$ is written as

$$f(x)\, dx, \tag{1}$$

and, from (i), the probability that an error parallel to OY lies between y and $y + dy$ is

$$f(y)\, dy. \tag{2}$$

The formula (1) expresses the probability that the falls of shot lie within the shaded vertical strip of width dx, and (2) expresses the probability that the falls of shot lie within the shaded horizontal strip of width dy.

The assumption (ii) and the principle of compound probability (§ 3·02) enable us to state that the probability that a shot will have errors between x and $x+dx$ and between y and $y+dy$ is

$$f(x)\,f(y)\,dx\,dy.$$

This is the probability that a shot will hit the small rectangle $dx\,dy$ at P. If α is any small area at P the probability that a shot will hit this area is then

$$\alpha f(x)\,f(y). \tag{3}$$

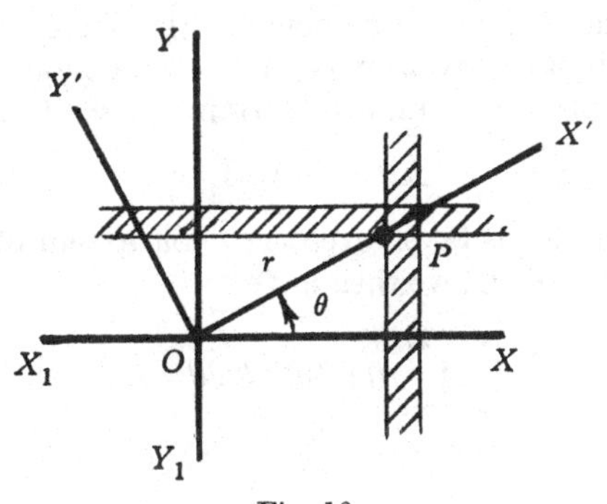

Fig. 10

Consider now axes OX', OY', the former passing through P. The polar coordinates of P with respect to the original axes are (r, θ) and the rectangular coordinates of P with respect to OX', OY' are $(r, 0)$; accordingly, by (3), the probability that a shot will hit the small area α at P is

$$\alpha f(r)\,f(0). \tag{4}$$

The two expressions (3) and (4) are equal; hence

$$f(x)\,f(y)=f(r)\,f(0). \tag{5}$$

The right-hand side of this equation is independent of θ; hence

$$f(y)\frac{\partial f(x)}{\partial\theta}+f(x)\frac{\partial f(y)}{\partial\theta}=0. \tag{6}$$

Now, $x=r\cos\theta$ and $y=r\sin\theta$; also

$$\frac{\partial f(x)}{\partial\theta}=\frac{\partial f(x)}{\partial x}\frac{\partial x}{\partial\theta}=-yf'(x) \quad \text{and} \quad \frac{\partial f(y)}{\partial\theta}=xf'(y);$$

hence (6) becomes
$$\frac{f'(x)}{xf(x)}=\frac{f'(y)}{yf(y)}.$$

Since x and y are independent, each side of this last equation must be equal to a constant, say, k. We have at once, on integration,

$$f(x)=C\,e^{kx^2},\quad f(y)=C\,e^{ky^2}, \tag{7}$$

the constant of integration, C, in each case being the same, in accordance with assumption (i).

In (7) we replace k by $-h^2$, in accordance with the principle that the probability of making a large error is less than that of making a small error; accordingly, we write

$$f(x)=C\,e^{-h^2x^2},\quad f(y)=C\,e^{-h^2y^2}. \tag{8}$$

Now, a shot must fall at some point on the plane and the sum, S, of all the probabilities, given by (4), for falls of shot wherever these may be on the plane amounts to 'certainty', so that $S=1$; hence, from (4),

$$\Sigma\alpha f(r)\,f(0)=1,$$

where the summation is taken over the whole extent of the plane. If α is the polar area $r\,dr\,d\theta$, we then have

$$\int_0^\infty\int_0^{2\pi} f(r)\,f(0)\,r\,dr\,d\theta=1. \tag{9}$$

But, from (5) and (8),

$$f(r)\,f(0)=C^2\,e^{-h^2(x^2+y^2)}=C^2\,e^{-h^2r^2};$$

hence (9) becomes

$$2\pi C^2\int_0^\infty e^{-h^2r^2}\,r\,dr\equiv\frac{\pi C^2}{h^2}\int_0^\infty e^{-z}\,dz=1,$$

where $z=h^2r^2$; the z integral is 1; hence $C^2=h^2/\pi$, and the function $f(x)$ is given by

$$f(x)=\frac{h}{\sqrt{\pi}}\,e^{-h^2x^2}. \tag{10}$$

This is the normal or Gaussian function derived in §3·05.

The assumptions by which (10) has been derived must not be lost sight of. As regards the first, namely, that the law of errors for one direction in the plane is the same as that for the perpendicular direction, it would seem that this is entirely reasonable for, under the conditions of the experiment, one set of axes is not likely to be more fundamental than any other. The second assumption, namely, that the errors x are independent of the errors y, has been criticized in some quarters as unwarrantable; the preceding proof would, of course, have to be abandoned if it were found, for example, that large errors x were invariably accompanied by small errors y. But, in the problem as stated, it is clearly necessary to introduce one or more assumptions if a solution is to be achieved. The final justification of the normal law is to be found in the results of experience, namely, that a series of numerous and trustworthy observations is generally in conformity with the normal law as established.

3·07. The law of errors for a linear function

We consider a linear function, ϕ, of two variables X and Y given by

$$\phi = aX + bY,$$

where a and b are constants.

Let x and y denote the errors in X and Y respectively; these errors are assumed to be independent and to obey the normal law with h and k as the respective moduli of precision. The probability that an error in X lies between x and $x+dx$ is then $\frac{h}{\sqrt{\pi}} e^{-h^2x^2}\,dx$, and the probability that an error in Y lies between y and $y+dy$ is $\frac{k}{\sqrt{\pi}} e^{-k^2y^2}\,dy$.

Let P be the probability that the errors between x and $x+dx$ and between y and $y+dy$ occur concurrently; then, by the principle of compound probability (§3·02),

$$P = \frac{hk}{\pi} e^{-(h^2x^2+k^2y^2)}\,dx\,dy. \tag{1}$$

Let u be the error in ϕ corresponding to errors x in X and y in Y; then

$$u = ax + by. \tag{2}$$

For an *assigned* value of u, the errors x and y can take any values consistent with (2).

Define† an error v by the linear expression

$$v = by - ax. \tag{3}$$

Then, from (2) and (3),

$$x = \frac{1}{2a}(u-v), \quad y = \frac{1}{2b}(u+v);$$

from these we obtain

$$dx\,dy \equiv \frac{\partial(x,y)}{\partial(u,v)}\,du\,dv = \frac{1}{2ab}\,du\,dv \tag{4}$$

† Any linear expression for v in terms of x and y, other than (2), would be suitable, but less simple than (3).

and $$h^2x^2+k^2y^2=\left(\alpha^2-\frac{\beta^2}{\alpha^2}\right)u^2+\alpha^2\left(v-\frac{\beta}{\alpha^2}u\right)^2, \tag{5}$$

where $$\alpha^2=\frac{1}{4}\left(\frac{h^2}{a^2}+\frac{k^2}{b^2}\right),\quad \beta=\frac{1}{4}\left(\frac{h^2}{a^2}-\frac{k^2}{b^2}\right). \tag{6}$$

Denote the coefficient of u^2 in (5) by H^2; then it is easily found from (6) that

$$\frac{1}{H^2}=\frac{a^2}{h^2}+\frac{b^2}{k^2}. \tag{7}$$

Then (1) becomes, by (4) and (5),

$$P=\frac{hk}{2\pi ab}e^{-H^2u^2}\exp\left[-\alpha^2\left(v-\frac{\beta u}{\alpha^2}\right)^2\right]du\,dv.$$

This expression for P is the probability that errors between u and $u+du$ and between v and $v+dv$ occur; hence the probability, Q, that errors between the assigned values u and $u+du$ occur for all possible values of v is given by

$$Q=\frac{hk}{2\pi ab}e^{-H^2u^2}du\int_{-\infty}^{\infty}\exp\left[-\alpha^2\left(v-\frac{\beta u}{\alpha^2}\right)^2\right]dv.$$

Since u has an assigned value, the value of the integral is $\sqrt{\pi}/\alpha$, by 1·14 (20). Further,

$$\frac{hk}{2\alpha ab}=\frac{hk}{\sqrt{(a^2k^2+b^2h^2)}}=H,\quad \text{by (7)}.$$

Hence, the probability that ϕ has errors between u and $u+du$ is

$$\frac{H}{\sqrt{\pi}}e^{-H^2u^2}du. \tag{8}$$

We thus have the important result that the law of errors for the linear function $\phi=aX+bY$ of two variables, subject to errors with moduli of precision h and k, is the normal law with modulus of precision, H, given by

$$\frac{1}{H^2}=\frac{a^2}{h^2}+\frac{b^2}{k^2}.$$

Clearly, this result can be generalized for a linear function of any number of variables. Thus, if

$$\phi=a_1X_1+a_2X_2+\ldots+a_nX_n, \tag{9}$$

where $X_1.\ X_2, \ldots$ are independent variables whose errors follow the normal law with moduli $h_1, h_2, \ldots$, then the errors, u, of ϕ follow the normal law given by (8), the modulus of precision, H, being given by

$$\frac{1}{H^2}=\frac{a_1^2}{h_1^2}+\frac{a_2^2}{h_2^2}+\ldots+\frac{a_n^2}{h_n^2}. \tag{10}$$

3·08. The law of errors for a linear function (alternative proof)

As in the previous section we consider the linear function

$$\phi=aX+bY;$$

the errors x and y in the variables X and Y follow the normal law with moduli h and k.

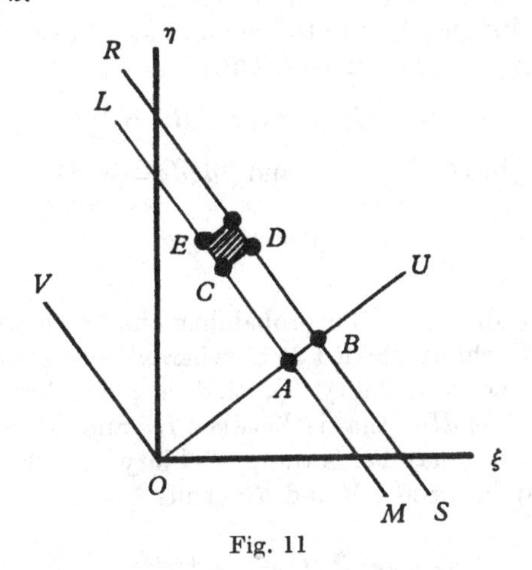

Fig. 11

From 3·07 (1), the probability, P, that independent errors between x and $x+dx$ and between y and $y+dy$ occur is given by

$$P=\frac{hk}{\pi}e^{-h^2x^2-k^2y^2}\,dx\,dy.$$

Let $$hx=\xi, \quad ky=\eta;$$

then $$P=\frac{1}{\pi}e^{-(\xi^2+\eta^2)}\,d\xi\,d\eta. \tag{1}$$

The error, u, in ϕ corresponding to the errors x and y is given by $u=ax+by$, or by

$$u=\frac{a}{h}\xi+\frac{b}{k}\eta. \tag{2}$$

Similarly, $$u+du=\frac{a}{h}(\xi+d\xi)+\frac{b}{k}(\eta+d\eta). \tag{3}$$

We regard (2) and (3) as the equations of the two parallel straight lines, LM and RS, in the ξ, η plane (Fig. 11). Take OU and OV to be new axes, OU being perpendicular to LM. Then, from (2), on applying the formula for the length of the perpendicular from the origin to a line,

$$OA = \frac{u}{\sqrt{\left(\frac{a^2}{h^2}+\frac{b^2}{k^2}\right)}} = Hu,$$

where
$$\frac{1}{H^2}=\frac{a^2}{h^2}+\frac{b^2}{k^2}. \tag{4}$$

Similarly, $OB = H(u+du)$; also, $AB = H\,du$.

Let C be a point on LM with the coordinates (ξ, η) and (U, V) with respect to the two systems of axes. Then

$$U \equiv OA = Hu, \quad dU \equiv AB = H\,du;$$

$$\xi^2+\eta^2 = U^2+V^2 \quad \text{and} \quad d\xi\,d\eta = dU\,dV.$$

Hence, from (1),
$$P = \frac{1}{\pi} e^{-(U^2+V^2)}\,dU\,dV.$$

Thus, geometrically, P is the probability that a point lies in the rectangle $dU\,dV$, shown shaded at C, where $dU = CD$ and $dV = CE$. Consequently, the probability, Q, that a point has coordinates between U and $U+dU$—that is, between Hu and $H(u+du)$—whatever the value of V may be, is the probability that the point lies in the infinite strip between LM and RS. Thus

$$Q = \frac{1}{\pi} e^{-U^2}\,dU \int_{-\infty}^{\infty} e^{-V^2}\,dV.$$

The value of the integral is $\sqrt{\pi}$. Then, since $U = Hu$,

$$Q = \frac{H}{\sqrt{\pi}} e^{-H^2u^2}\,du.$$

Thus the normal law is reproduced for a linear function, the modulus H being given by (4). The generalization, represented by 3·09 (9) and (10), follows.

CHAPTER 4

MEASURES OF PRECISION

4·01. Introduction

In this chapter we consider the problem of deriving, for practical purposes, formulae which enable us (i) to obtain the best or most probable value of an unknown quantity from a number, n, of observations, x_i, of equal reliability, and (ii) to assess the precision of this value. As regards (i) the value concerned is, according to the principle of least squares in § 2·11 and to Gauss's postulate in § 3·04, the arithmetic mean, a, which is readily calculated from the formula $a = \Sigma x_i/n$. As regards (ii) the formulae which are summarized later in § 4·11 are, in fact, a set of standardized rules derived on the assumption that the observational errors are accidental, either in conformity with Hagen's hypothesis (§ 2·17) or subject to the play of probability as in Chapter 3. In seeking a set of rules, two considerations are important.

First, the theoretical basis of the rules must be reasonable and, in particular, the postulate or rule which asserts that the arithmetic mean is the best or most probable value of the unknown must be justified in the light of experience. Now, the basis of the rules to be derived is the normal law—developed in the two previous chapters partly out of *a priori* considerations—which, in an ideal set of observations, appears to be not without significance. The justification of the rule concerning the arithmetic mean is that in several series of observations made with the greatest care and with efficient apparatus the arithmetic means for the individual series are found to be substantially consistent; this is true notably in many classes of astronomical observations. Further, when a large number of observations of an unknown quantity are analysed, it is found that the errors, these being identified with the residuals, follow remarkably closely the error law, as illustrated in § 2·20. At the worst, the normal law, when shed of its theoretical origins, may be regarded as a standard with which the best observations, if sufficiently numerous, are generally found to conform.

Secondly, the rules should be simple and easy of numerical application; as we shall see, the normal law has conspicuous merits in this connexion. Further, the rules should be accepted in all quarters as the *standard rules*.

The rules are primarily based on the assumption that the observations are sufficiently numerous and, in this case, when the errors follow the normal law, the rules are explicit and worthy of confidence. But, when the number of observations is small, say 6 or 7, the assessment of the precision of the arithmetic mean according to the rules can be regarded only as approximate; in such a case the results which we obtain are the best that can be achieved under the circumstances. In § 4·16 the problem of assessing this degree of approximation will be discussed.

4·02. The normal law

We denote an error of observation by ϵ. According to 2·17 (11), the expression of the normal law is as follows. If the number, N, of observations is large the frequency of errors between ϵ and $\epsilon+d\epsilon$ is $Nf(\epsilon)\,d\epsilon$, where

$$f(\epsilon)=\frac{h}{\sqrt{\pi}}e^{-h^2\epsilon^2}; \tag{1}$$

if this frequency is denoted by $n(\epsilon)\,d\epsilon$, then

$$n(\epsilon)\,d\epsilon=Nf(\epsilon)\,d\epsilon. \tag{2}$$

Also, the relative frequency of errors between ϵ and $d\epsilon$ is $f(\epsilon)\,d\epsilon$.

According to Chapter 3 the normal law, derived from the concepts of probability, takes the following form: the probability of an error between ϵ and $\epsilon+d\epsilon$ is $f(\epsilon)\,d\epsilon$. The empirical definition of probability (§ 3·01) enables us to pass to frequency so that, if N is large, the probable frequency, $n(\epsilon)\,d\epsilon$, of errors between ϵ and $\epsilon+d\epsilon$ is also given by (2).

In (1), h is the modulus of precision and, in a series of observations whose errors follow the normal law, the more accurate the observations the larger is h. But h, being a theoretical quantity depending on the observational errors which we do not know, is ill-adapted as a *practical* measure of precision, and in the next sections we seek other means of ascertaining the precision of a series of observations in terms of the observed quantities themselves.

4·03. The formula $\mu=\dfrac{1}{h\sqrt{2}}$ for a continuous distribution

When the errors follow the normal law and form a continuous distribution, the frequency $n(\epsilon)\,d\epsilon$ of errors between ϵ and $\epsilon+d\epsilon$ is given by 4·02 (2), namely,

$$n(\epsilon)\,d\epsilon=Nf(\epsilon)\,d\epsilon,$$

in which N is supposed to be a large number. Since the square of each of these errors can be taken to be ϵ^2, then

$$N\mu^2 = \Sigma n(\epsilon)\, d\epsilon \,.\, \epsilon^2, \tag{1}$$

where the summation is taken for all values of ϵ between $-\infty$ and $+\infty$. Then (1) is written as

$$N\mu^2 = N\int_{-\infty}^{\infty} \epsilon^2 f(\epsilon)\, d\epsilon,$$

or, by means of 4·02 (1),

$$\mu^2 = \frac{h}{\sqrt{\pi}}\int_{-\infty}^{\infty} \epsilon^2 e^{-h^2\epsilon^2}\, d\epsilon.$$

By 1·14 (21), with $m=1$, the value of the integral is $\sqrt{\pi}/(2h^3)$; hence

$$\mu^2 = \frac{1}{2h^2}, \tag{2}$$

or, since μ and h are positive quantities,

$$\mu = \frac{1}{h\sqrt{2}}. \tag{3}$$

It is to be remarked that μ, like h, is a theoretical quantity associated with the normal law.

4·04. The formula $\mu = \dfrac{1}{h\sqrt{2}}$ for discrete observations

Consider n discrete independent observations $x_1, x_2, \ldots, x_n$, of equal reliability, of a true quantity x; further, suppose that c is a small quantity related to the degree of accuracy attainable by the measuring instrument concerned. Then, by the application of the normal law in terms of probability, the probability that the error ϵ_i of the measure x_i lies between $\epsilon_i - \frac{1}{2}c$ and $\epsilon_i + \frac{1}{2}c$ is $\dfrac{ch}{\sqrt{\pi}} e^{-h^2\epsilon_i^2}$, where h is the modulus of precision at present unknown. The assumption of equal reliability implies that h is the same for all the observations. Hence, the probability, P, that all the errors occur is given, according to the principle of compound probability in § 3·02, by

$$P = Ch^n e^{-h^2\Sigma\epsilon_i^2}$$

where $C = \{c/\sqrt{\pi}\}^n$, or by $\quad P = Ch^n e^{-nh^2\mu^2}, \qquad (1)$

where μ is the standard error defined by $n\mu^2 = \Sigma\epsilon_i^2$.

With the observations actually made, the errors ϵ_i, although unknown, are defined implicitly, and so $\Sigma\epsilon_i^2$, or μ^2 in (1), may be regarded as representing a specific property of the series of observations.

The most probable value of h is such that P is a maximum, or that $\log P$ is a maximum. The first condition for a maximum is

$$\frac{d}{dh}(\log P) \equiv \frac{d}{dh}(n \log h - nh^2\mu^2) = 0,$$

whence
$$\mu^2 = \frac{1}{2h^2} \tag{2}$$

or
$$\mu = \frac{1}{h\sqrt{2}}. \tag{3}$$

Also,
$$\frac{d^2}{dh^2}(\log P) = -\frac{n}{h^2} - 2n\mu^2;$$

accordingly, since the expression on the right is negative, the second condition for a maximum is satisfied.

The formulae (2) and (3) are the same as those derived in the previous section for a continuous distribution.

4·05. The formula for μ in terms of the squares of the residuals

Let v_1 denote the residual for an observation x_1 and a the arithmetic mean of n observations, x_i. Then

$$v_1 \equiv x_1 - a = x_1 - \frac{1}{n}(x_1 + x_2 + \ldots + x_n)$$

or
$$v_1 = \frac{n-1}{n} x_1 - \frac{x_2}{n} - \frac{x_3}{n} - \ldots - \frac{x_n}{n}. \tag{1}$$

Thus v_1 is a linear function of the independent quantities $x_1, x_2, \ldots, x_n$. Assuming that the observations are equally reliable, let h be the modulus of precision associated with each x_i so that the standard error, μ, is given, as in § 4·04, by

$$\mu^2 = \frac{1}{2h^2}. \tag{2}$$

Also, let H be the modulus of precision of the residual v_1; then, by application of 3·07 (10), we obtain from (1),

$$\frac{1}{H^2} = \left(\frac{n-1}{n}\right)^2 \frac{1}{h^2} + (n-1)\frac{1}{n^2h^2} = \frac{n-1}{n}\frac{1}{h^2}. \tag{3}$$

This is clearly true for any residual, that is to say, the modulus H is the same for all residuals.

If μ_1^2 is the mean of the squares of the residuals, then

$$n\mu_1^2 = \Sigma v_i^2. \tag{4}$$

Also, since H is the modulus associated with the v's then, by applying (2) and by means of (3), we have

$$\mu_1^2=\frac{1}{2H^2}=\frac{n-1}{n}\frac{1}{2h^2},$$

from which, by (4) and (2), $\mu^2=\dfrac{\Sigma v_i^2}{n-1}$.

In the Gaussian notation, the last equation is

$$\mu^2=\frac{[vv]}{n-1} \tag{5}$$

or

$$\mu=\sqrt{\frac{[vv]}{n-1}}. \tag{6}$$

The formula (5) is that derived in §2·13 from elementary considerations.

The value of the arithmetic mean, a, is easily calculated from the measures $x_1, x_2, \ldots, x_n$, and from it the residuals, v_i, can be written down. The formula (6) then enables us to calculate the value of μ.

4·06. The standard error of the arithmetic mean

The arithmetic mean, a, is given by

$$a=\frac{x_1}{n}+\frac{x_2}{n}+\ldots+\frac{x_n}{n}.$$

As before, h is the modulus associated with each x_i. Let H_a be the modulus associated with a. Then, since $x_1, x_2, \ldots$ are independent, we obtain, by the theorem in §3·07,

$$\frac{1}{H_a^2}=n\left(\frac{1}{n}\right)^2\frac{1}{h^2}=\frac{1}{nh^2}. \tag{1}$$

If μ_a is the standard error corresponding to H_a, then

$$\frac{1}{H_a^2}=2\mu_a^2. \tag{2}$$

But, since $\mu^2=1/(2h^2)$, we obtain from (1) and (2)

$$\mu_a^2=\frac{\mu^2}{n}, \tag{3}$$

or, by means of 4·05 (5), $\mu_a^2=\dfrac{[vv]}{n(n-1)}$. (4)

The solution is written

$$x = a \pm \mu_a \quad (\text{S.E.}), \tag{5}$$

a being the most probable value of the unknown and μ_a the standard error of a; the attachment of the double sign to μ_a in (5) follows from the arguments in §§ 2·12 and 2·14.

4·07. The formula $\eta = \dfrac{1}{h\sqrt{\pi}}$

We assume as in § 4·02 that the number, N, of observations is large and that the errors, ϵ, follow the normal law.

Let η denote the mean of the errors taken *without regard to sign* so that $\eta = \frac{1}{n}\Sigma\,|\,\epsilon_i\,|$; we refer to η as the *average error*.

Since the error function is symmetrical, we need only consider the positive errors of which the number is $\frac{1}{2}N$.

As before, let $n(\epsilon)\,d\epsilon$ denote the frequency of errors between ϵ and $\epsilon + d\epsilon$. Then the definition of η gives

$$\tfrac{1}{2}N\eta = \Sigma n(\epsilon)\,d\epsilon\,.\,\epsilon$$

$$= N\Sigma f(\epsilon)\,d\epsilon\,.\,\epsilon \quad \text{by } 4{\cdot}02\,(2),$$

where the summation is taken for all values of ϵ from 0 to $+\infty$. Hence we have

$$\eta = \frac{2h}{\sqrt{\pi}} \int_0^\infty \epsilon\, e^{-h^2\epsilon^2}\,d\epsilon.$$

The value of the integral is $1/(2h^2)$. Hence

$$\eta = \frac{1}{h\sqrt{\pi}}. \tag{1}$$

This formula gives the *average error*, η, in terms of the modulus h when, in the ideal case, the number of observations is large and the errors are distributed according to the normal law; also, like h and μ, η is a theoretical quantity.

4·08. Peters's formulae for μ and μ_a in terms of the numerical values of the residuals

Let η_1 denote the *average value* of the residuals v_i of n observations x_i. Then $v_i = x_i - a$ and

$$\eta_1 = \frac{1}{n}\Sigma\,|\,v_i\,|. \tag{1}$$

Let H be the modulus of precision associated with the residuals; then, from 4·05 (3),

$$\frac{1}{H^2}=\frac{n-1}{n}\frac{1}{h^2}. \tag{2}$$

If the observations are supposed to be sufficiently numerous, we apply the formula 4·07 (1) for a continuous distribution so that, in the present case,

$$\eta_1=\frac{1}{H\sqrt{\pi}}.$$

Hence, from (1) and (2),

$$\frac{1}{n}\Sigma\,|\,v_i\,|=\frac{1}{h}\sqrt{\left(\frac{n-1}{\pi n}\right)}. \tag{3}$$

It is convenient to write $\Sigma\,|\,v_i\,|\equiv[v]$ in which, of course, only *numerical values of the residuals* are involved. Then (3) becomes, on rearrangement,

$$\frac{1}{h}=\frac{\sqrt{\pi}\,[v]}{\sqrt{\{n(n-1)\}}}.$$

But, $1/h=\mu\sqrt{2}$; hence

$$\mu=\sqrt{\left(\frac{\pi}{2}\right)}\frac{[v]}{\sqrt{\{n(n-1)\}}}. \tag{4}$$

Replacing $\sqrt{(\frac{1}{2}\pi)}$ by its numerical value, we have

$$\mu=1{\cdot}2533\frac{[v]}{\sqrt{\{n(n-1)\}}}. \tag{5}$$

This is Peters's formula for the calculation of the standard error, μ.

The formula (5) is much simpler, as regards computation, than the formula

$$\mu=\sqrt{\left\{\frac{[vv]}{n-1}\right\}} \tag{6}$$

obtained in §4·05; as we shall see in §4·16 there is little to choose between them from the point of view of the assessment of the accuracy of the standard error, μ.

Since the standard error, μ_a, of the arithmetic mean is $\mu/\sqrt{n}$, Peters's formula for μ_a is written

$$\mu_a=\sqrt{\left(\frac{\pi}{2}\right)}\frac{[v]}{n\sqrt{(n-1)}}. \tag{7}$$

As regards the calculation of μ and μ_a in any particular problem, we regard (4) and (7) as alternative formulae to (6) and the corresponding formula for μ_a, namely,

$$\mu_a=\sqrt{\left\{\frac{[vv]}{n(n-1)}\right\}}.$$

4·09. Probable error

So far the basic practical measure of precision introduced for a series of observations is the standard error μ. A second measure of precision, inaptly† named *probable error* and denoted by r, is defined as follows. Let N be the number of observations the errors of which obey the normal law, N being large. Let AB and CD in Fig. 12 be the ordinates of the normal curve corresponding to errors $+r$ and $-r$; then r is defined to be such that the total number of errors in the range $-r \leqslant \epsilon \leqslant r$ is $\frac{1}{2}N$, or, alternatively, that the relative frequency in this

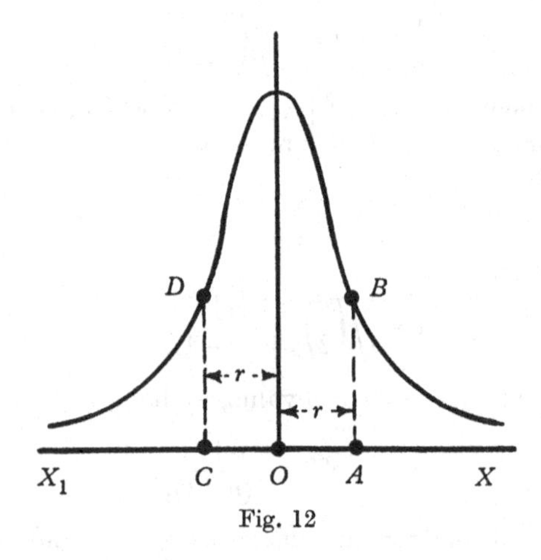

Fig. 12

range is $\frac{1}{2}$. It will be noticed that OA and OC (or $+r$ and $-r$) correspond to the *quartiles* defined in § 1·04 (iii). Now the relative frequency in the range is

$$\frac{2h}{\sqrt{\pi}}\int_0^r e^{-h^2\epsilon^2}\,d\epsilon, \quad \text{or if } t=h\epsilon, \quad \frac{2}{\sqrt{\pi}}\int_0^{hr} e^{-t^2}\,dt,$$

that is, by 1·15 (10), erf (hr). Hence r is given by

$$\operatorname{erf}(hr)=\tfrac{1}{2}.$$

The table of the erf function (Appendix 1) shows that

$$hr=0{\cdot}4769. \tag{1}$$

† The most probable value of one of a large number of errors following the error law is $\frac{h}{\sqrt{\pi}}\int_{-\infty}^{\infty} \epsilon e^{-h^2\epsilon^2}\,d\epsilon$ or zero, and the most probable value of the errors taken without regard to sign is η or $1/(h\sqrt{\pi})$.

It is convenient to write (1) as

$$hr=\rho, \tag{2}$$

where
$$\rho=0{\cdot}4769. \tag{3}$$

Then
$$r=\rho\frac{1}{h}. \tag{4}$$

Since $\mu=1/(h\sqrt{2})$, this last formula enables us to connect r with μ. Thus $r=\rho\mu\sqrt{2}$ or

$$r=0{\cdot}6745\mu. \tag{5}$$

In the preceding sections we have derived formulae for μ in terms of $[vv]$ and of $[v]$. From (5) we then have the corresponding formulae for r:

$$r=0{\cdot}6745\sqrt{\left\{\frac{[vv]}{n-1}\right\}} \tag{6}$$

and
$$r=0{\cdot}6745\sqrt{\left(\frac{\pi}{2}\right)}\frac{[v]}{\sqrt{\{n(n-1)\}}},$$

or, on inserting the numerical value of $\sqrt{(\frac{1}{2}\pi)}$,

$$r=0{\cdot}8453\frac{[v]}{\sqrt{\{n(n-1)\}}}. \tag{7}$$

The *probable error of the arithmetic mean*, denoted by r_a, is given, from (5), by

$$r_a=0{\cdot}6745\mu_a,$$

leading to the two alternative formulae

$$r_a=0{\cdot}6745\sqrt{\left\{\frac{[vv]}{n(n-1)}\right\}}$$

and
$$r_a=0{\cdot}8453\frac{[v]}{n\sqrt{(n-1)}}.$$

Note. In numerical computations it is sometimes sufficient to take the approximations

$0{\cdot}6745 \simeq 2/3$, or, more accurately, $27/40$,

and $0{\cdot}8453 \simeq 5/6$, or, more accurately, $17/20$.

4·10. Remarks on measures of precision

For reference we collect the various formulae for μ, η and r in terms of h; they are

$$\mu=\frac{1}{h\sqrt{2}},\quad \eta=\frac{1}{h\sqrt{\pi}},\quad r=\frac{\rho}{h}, \tag{1}$$

where $\rho = 0{\cdot}4769$. In particular we write

$$r = 0{\cdot}6745\mu = 0{\cdot}8453\eta \tag{2}$$

and

$$\mu = 1{\cdot}4826r = 1{\cdot}2533\eta. \tag{3}$$

As we have mentioned in §2·19, the greater the precision of the observations, the greater is the value of h; hence, from (1), the greater the precision of the observations, the smaller are the values of μ, η and r.

The assessment of the accuracy of a series of observations is made, in practice, either in terms of μ or in terms of r.

In many European countries it is usual to employ μ and μ_a for the criteria of accuracy; in English-speaking countries the use of r and r_a is favoured.

The transformation from r to μ is effected by means of (3), and from μ to r by means of (2).

In the examples which follow after the next section, the results will be given in terms of both standard error and probable error, these being indicated by S.E. and P.E. respectively.

4·11. Summary of rules

It is convenient for reference to gather together the principal formulae used in the statistical treatment of equations in one unknown; it is assumed that the n measured quantities x_i are equally reliable.

First, the most probable value of the unknown is the arithmetic mean.

Secondly, the characteristic distribution of the errors of observation is expressed either in terms of the standard error, μ, or in terms of the probable error, r; the accuracy of the arithmetic mean is assessed either in terms of μ_a (S.E.) or in terms of r_a (P.E.).

There are two methods of computing μ and r; the first is by finding the sum of the squares of the residuals, that is, $[vv]$; the second is by finding the sum of the residuals taken without regard to sign, that is, $[v]$.

Method I:

$$\mu = \sqrt{\left\{\frac{[vv]}{n-1}\right\}},$$

$$r = 0{\cdot}6745\sqrt{\left\{\frac{[vv]}{n-1}\right\}},$$

$$\mu_a = \sqrt{\left\{\frac{[vv]}{n(n-1)}\right\}},$$

$$r_a = 0{\cdot}6745\sqrt{\left\{\frac{[vv]}{n(n-1)}\right\}}.$$

Method II:

$$\mu = 1{\cdot}2533 \frac{[v]}{\sqrt{\{n(n-1)\}}},$$

$$r = 0{\cdot}8453 \frac{[v]}{\sqrt{\{n(n-1)\}}},$$

$$\mu_a = 1{\cdot}2533 \frac{[v]}{n\sqrt{(n-1)}},$$

$$r_a = 0{\cdot}8453 \frac{[v]}{n\sqrt{(n-1)}}.$$

Also, for reference,

$$r = 0{\cdot}6745\mu,$$

$$\mu_a = \frac{\mu}{\sqrt{n}},$$

$$r_a = \frac{r}{\sqrt{n}}.$$

4·12. Example 1: the position angle of a double star

The example relates to seven observations of the position angle† θ of the double star β 1077 made in April 1937 (*Greenwich Observations*, 1937, B. 11).

The values of θ for the seven results published range from 301°·9 to 312°·9; we write

$$\theta = 300° + x°.$$

The values of x are given in the first column of Table 8.

Table 8. *Measures of position angle*

x	v +	v −	v^2
1·9	—	6·3	39·7
2·2	—	6·0	36·0
11·8	3·6	—	13·0
9·5	1·3	—	1·7
9·9	1·7	—	2·9
12·9	4·7	—	22·1
9·5	1·3	—	1·7
57·7	**12·6**	**12·3**	**117·1**

† The position angle with reference to two stars, A and B, close together in the sky, is the angle which AB makes with the direction of the north celestial pole and is measured from 0° to 360°; as the separation AB is frequently of the order of 1″, the measurement of θ requires considerable skill and experience on the part of the observer.

The sum of the first column is 57·7, from which the arithmetical mean, a, of the seven values of x is 8·2. The residuals and their squares are in the remaining columns.

The check that Σv should be equal to zero is satisfied very nearly, the slight discrepancy being due to the omission of the second decimal place in taking the mean to be 8·2 instead of 8·24.

From Table 8 we have:

$$n=7; \quad [vv]=117{\cdot}1; \quad [v]=12{\cdot}6+12{\cdot}3=24{\cdot}9.$$

We apply the rules as summarized in §4·11.

	Method I	Method II
μ	$\sqrt{\dfrac{117{\cdot}1}{6}}=4{\cdot}4$	$\dfrac{1{\cdot}2533\times 24{\cdot}9}{\sqrt{42}}=4{\cdot}8$
μ_a	$\dfrac{\mu}{\sqrt{7}}=1{\cdot}7$	$\dfrac{\mu}{\sqrt{7}}=1{\cdot}8$
r	$0{\cdot}6745\mu=3{\cdot}0$	$0{\cdot}6745\mu=3{\cdot}2$
r_a	$\dfrac{r}{\sqrt{7}}=1{\cdot}1$	$\dfrac{r}{\sqrt{7}}=1{\cdot}2$

When one considers the comparatively small number of observations and the relatively high dispersion in the values of θ, the agreement between the results by the two methods of computing μ, etc., is satisfactory.

Taking the larger values of μ, etc., by Method II we write the definitive solutions first in terms of standard error and secondly in terms of probable error as follows:

$$\theta \equiv 308^\circ{\cdot}2 \pm \mu_a = 308^\circ{\cdot}2 \pm 1^\circ{\cdot}8 \quad (\text{S.E.}),$$

and

$$\theta \equiv 308^\circ{\cdot}2 \pm r_a = 308^\circ{\cdot}2 \pm 1^\circ{\cdot}2 \quad (\text{P.E.}).$$

4·13. Example 2: the mechanical equivalent of heat

We denote this physical constant by X (ergs per 20° C. calorie). It is found that the measures of X cluster around the value $4{\cdot}1800\times 10^7$ or $41{,}800\times 10^3$; it is convenient to refer to X in terms of the unit 10^3. Further, it is advantageous to express X, in terms of this unit, by

$$X=41{,}800+x. \tag{1}$$

From nine measures† of X the corresponding values of x are shown in the first column of Table 9.

† F. E. Hoare, *A Textbook of Thermodynamics*, 2nd ed. (Arnold, 1938), p. 11.

Table 9. *Measures of the mechanical equivalent of heat*

x		v		
+	−	+	−	v^2
—	120	—	126	15,876
22	—	16	—	256
—	42	—	48	2,304
9	—	3	—	9
104	—	98	—	9,604
98	—	92	—	8,464
—	5	—	11	121
—	33	—	39	1,521
21	—	15	—	225
254	**200**	**224**	**224**	**38,380**

The arithmetic mean, a, of the nine values of x is given by

$$a = \tfrac{1}{9}(254 - 200) = 6,$$

from which the residuals v ($\equiv x - a$) are derived. The best or most probable value of X is thus, from (1), $41{,}806 \times 10^3$. Also, $\Sigma v = 0$.

From the table we have

$$n = 9; \quad [vv] = 38{,}380; \quad [v] = 448.$$

The details of the calculation of μ, etc., follow:

	Method I		Method II	
μ	$\sqrt{\left(\dfrac{38380}{8}\right)}$	$=69{\cdot}3$	$\dfrac{1{\cdot}2533 \times 448}{\sqrt{72}}$	$=66{\cdot}2$
μ_a	$\dfrac{\mu}{\sqrt{9}}$	$=23{\cdot}1$	$\dfrac{\mu}{\sqrt{9}}$	$=22{\cdot}1$
r	$0{\cdot}6745\mu$	$=46{\cdot}7$	$0{\cdot}6745\mu$	$=44{\cdot}6$
r_a	$\dfrac{r}{\sqrt{9}}$	$=15{\cdot}6$	$\dfrac{r}{\sqrt{9}}$	$=14{\cdot}9$

If we take the mean of the two values of μ_a and the mean of the two values of r_a, derived by the two methods, the solutions are, to the nearest integer,

$$X = (41{,}806 \pm 23) \times 10^3 \quad \text{(S.E.)},$$

and

$$X = (41{,}806 \pm 15) \times 10^3 \quad \text{(P.E.)}.$$

4·14. The standard error and probable error of a linear function of independent variables

As in § 2·16 we consider a linear function, ϕ, of the independent variables x and y given by

$$\phi = ax + by,$$

where a and b are constants. It is supposed that several measures of x, for example, have been made, from which the mean, $\bar{x}$, and the standard error, μ_x, have been calculated according to the rules stated in §4·11; since, in general,

$$h^2 = 1/(2\mu^2), \tag{1}$$

the modulus, h_x, for the measures of x is expressed in terms of μ_x.

Similarly, the values of $\bar{y}$, μ_y and h_y are obtained, the number of measures of y being not necessarily the same as for x.

The most probable value of ϕ is given by its mean, namely,

$$\phi = a\bar{x} + b\bar{y}.$$

Let μ_ϕ denote the standard error of ϕ, with H as the corresponding modulus. Now, the variables are independent; hence, by 3·07 (10),

$$\frac{1}{H^2} = \frac{a^2}{h_x^2} + \frac{b^2}{h_y^2},$$

so that, by (1), $$\mu_\phi^2 = a^2\mu_x^2 + b^2\mu_y^2 \quad \text{(S.E.)}. \tag{2}$$

Since, in general, $r = 0{\cdot}6745\mu$, the formula (2) can be expressed in terms of probable errors as

$$r_\phi^2 = a^2 r_x^2 + b^2 r_y^2 \quad \text{(P.E.)}, \tag{3}$$

where r_ϕ is the probable error of ϕ, and r_x and r_y are the probable errors of the measures of x and y.

It is to be noted that the formula (2) is the same as 2·16 (3).

In §2·16 it was assumed that the number of measures of x was the same as the number for y, and that the measure x_i was associated with the measure y_i. It is now seen that when the principles of probability are invoked, these restrictions are no longer necessary so far as the derivation of (2) and of (3) is concerned.

The formula (2) and (3) can clearly be generalized for a linear function of any number of independent variables. Thus, for example, if

$$\phi = ax + by + cz,$$

then $$\mu_\phi^2 = a^2\mu_x^2 + b^2\mu_y^2 + c^2\mu_z^2 \quad \text{(S.E.)}. \tag{4}$$

Similarly, $$r_\phi^2 = a^2 r_x^2 + b^2 r_y^2 + c^2 r_z^2 \quad \text{(P.E.)}.$$

The theorem can also be expressed in terms of errors. If $\phi(x, y)$, x and y are true values the error, e, of $\phi(\bar{x}, \bar{y})$ is given by

$$e = \phi(\bar{x}, \bar{y}) - \phi(x, y) = a(\bar{x} - x) + b(\bar{y} - y).$$

Then, if ϵ_x and ϵ_y denote the errors of $\bar{x}$ and $\bar{y}$,

$$\epsilon_x = \bar{x} - x, \quad \epsilon_y = \bar{y} - y;$$

hence $$e = a\epsilon_x + b\epsilon_y,$$
from which, as before,
$$\mu_e^2 \equiv \mu_\phi^2 = a^2\mu_x^2 + b^2\mu_y^2.$$

Note. If the function, ϕ, in two variables x and y contains also a constant so that
$$\phi = ax + by + k,$$
we can write this as $$\phi = ax + by + kz,$$
where $z = 1$. Now, z being constant, the 'standard error', μ_z, is zero; hence, by (4),
$$\mu_\phi^2 = a^2\mu_x^2 + b^2\mu_y^2,$$
as in (2). The formula (3) follows.

4·15. Illustrative example

We consider the simple problems concerned with the measurement of (i) the perimeter of a rectangular field and (ii) the area.

Let x and y denote the lengths of two adjacent sides of the field and we suppose that m measures of x are made and, independently, n measures of y are made.

If a denotes the arithmetic mean of the measures x_i $(i = 1, 2, \ldots, m)$ and u_i is the corresponding residual, then
$$u_i = x_i - a \quad \text{and} \quad \mu_x^2 = \frac{[uu]}{m-1}.$$
Thus, μ_x is computed.

Similarly, if b denotes the arithmetic mean of the measures y_j $(j = 1, 2, \ldots, n)$ and v_j is the corresponding residual, then
$$v_j = y_j - b \quad \text{and} \quad \mu_y^2 = \frac{[vv]}{n-1}.$$

(i) The perimeter, P, is given by $P = 2x + 2y$; hence by 4·14 (2),
$$\mu_p^2 = 4(\mu_x^2 + \mu_y^2),$$
and, by 4·14 (3), $$r_p^2 = 4(r_x^2 + r_y^2).$$

(ii) The area, A, is given by $A = xy$. This expression for A is not linear. But write
$$x = a + \xi, \quad y = b + \eta; \tag{1}$$
then $$A = ab + b\xi + a\eta + \xi\eta.$$

Now, in a series of reasonably precise measures the final term, $\xi\eta$, may be expected to be small compared with the remaining terms. Neglecting $\xi\eta$, we then have, by 4·14 (2),
$$\mu_A^2 = b^2\mu_\xi^2 + a^2\mu_\eta^2.$$

But, by the *Note* at the end of the previous section we have, by (1),

$$\mu_x = \mu_\xi, \quad \mu_y = \mu_\eta.$$

Hence
$$\mu_A^2 = b^2\mu_x^2 + a^2\mu_y^2.$$

Similarly,
$$r_A^2 = b^2 r_x^2 + a^2 r_y^2.$$

4·16. Precision of the standard and probable errors

(i) *Introduction.* We shall suppose that n measures of a quantity, x, are made; these may be regarded as forming a *sample* of a very large number, N, of measures, actual or potential, which substantially constitute a continuous distribution. We shall assume that all measures, whether n or N in number, are considered equally reliable so that the associated modulus, h, is the same for all. If there had been N measures, as for a continuous distribution, the standard error, μ, would be given accurately by

$$\mu^2 = \frac{1}{2h^2}. \tag{1}$$

We regard μ as the true standard error.

The two formulae which have been derived to enable us to calculate the standard error of n actual measures, denoted here for convenience by μ_c, are

$$\mu_c^2 = \frac{\Sigma v_i^2}{n-1}, \tag{2}$$

and (Peters's formula)

$$\mu_c = \sqrt{\frac{\pi}{2}} \frac{\Sigma |v_i|}{\sqrt{\{n(n-1)\}}}. \tag{3}$$

By 4·05 (1) the residual, v_1, for example, is given by

$$v_1 = \frac{n-1}{n} x_1 - \frac{x_2}{n} - \ldots - \frac{x_n}{n},$$

and, if H is the modulus associated with the residual, then, by 4·05 (3),

$$\frac{1}{H^2} = \frac{n-1}{n} \frac{1}{h^2},$$

and, by (1),
$$\frac{1}{H^2} = \frac{2(n-1)\mu^2}{n}. \tag{4}$$

The modulus, H, is the same for all residuals.

If n is a small number, say 8, the formulae (2) and (3) may be expected to yield only approximate values of the true standard error, μ.

The object of this section is to try to assess the error inherent in the computed value, μ_c, derived from the n measures of the sample; we deal with (2) and (3) separately.

(ii) *Formula for μ_c in terms of Σv_i^2.* We consider formula (2). Let e denote the error of μ_c^2 so that

$$e = \mu_c^2 - \mu^2. \tag{5}$$

This error, which is related to the sample of n measures, may be positive or negative. From (5)

$$(e+\mu^2)^2 = \mu_c^4,$$

or, by (2),

$$(n-1)^2 e^2 + 2(n-1)^2 \mu^2 e = \{\Sigma v_i^2\} - (n-1)^2 \mu^4$$

$$= \Sigma v_i^4 + \Sigma v_i^2 (\Sigma v_j^2) - (n-1)^2 \mu^4 \quad (j \neq i).$$

Let $M(X)$ denote the mean value of X; in particular, let $M(e^2) = \mu_0^2$, so that μ_0 is the standard error of the e's. Then

$$(n-1)^2 \mu_0^2 + 2(n-1)^2 \mu^2 M(e) = \Sigma M(v_i^4) + \Sigma M(v_i^2) . \Sigma M(v_j^2) - (n-1)^2 \mu^4. \tag{6}$$

Now, in general, if k is the modulus associated with a quantity, x, subject to the normal law, then

$$M(x^m) = \frac{k}{\sqrt{\pi}} \int_{-\infty}^{\infty} x^m e^{-k^2 x^2} dx. \tag{7}$$

From the values of the integrals in § 1·14 (v), we have, applying (7) and remembering that the modulus associated with v is H, the following:

$$m=1, \quad M(e) = 0;$$

$$m=2, \quad M(v^2) = \frac{H}{\sqrt{\pi}} \int_{-\infty}^{\infty} v^2 e^{-H^2 v^2} dv = \frac{1}{2H^2};$$

$$m=4, \quad M(v^4) = \frac{H}{\sqrt{\pi}} \int_{-\infty}^{\infty} v^4 e^{-H^2 v^2} dv = \frac{3}{4H^4}.$$

Then, since there are n terms in the i summations and $n-1$ terms in the j summation, (6) becomes

$$(n-1)^2 \mu_0^2 = \frac{3n}{4H^4} + n\left(\frac{1}{2H^2}\right)(n-1)\left(\frac{1}{2H^2}\right) - (n-1)^2 \mu^4,$$

or, by means of (4), $$\mu_0^2 = \frac{2}{n}\mu^4,$$

or $$\mu_0 = \mu^2\sqrt{\frac{2}{n}},$$

which thus gives the standard error of e. Hence, in terms of standard error, we have

$$e \equiv \mu_c^2 - \mu^2 = \pm\mu^2\sqrt{\frac{2}{n}} \quad \text{(S.E.)}. \tag{8}$$

If r_0 is the probable error corresponding to μ_0, then

$$r_0 = \alpha\mu_0 = \alpha\mu^2\sqrt{\frac{2}{n}},$$

where $\alpha = 0{\cdot}6745$. Then, in terms of probable error, (8) becomes

$$\mu_c^2 - \mu^2 = \pm\alpha\mu^2\sqrt{\frac{2}{n}} \quad \text{(P.E.)},$$

or, if r and r_c are respectively the true probable error and the probable error calculated for the sample so that $r = \alpha\mu$ and $r_c = \alpha\mu_c$, then

$$r_c^2 - r^2 = \pm\alpha r^2\sqrt{\frac{2}{n}} \quad \text{(P.E.)}. \tag{9}$$

From (8), $$\frac{\mu_c}{\mu} = \left(1 \pm \sqrt{\frac{2}{n}}\right)^{\frac{1}{2}};$$

if n is not too small, say $n \not< 6$, we obtain an approximation to the right-hand side, by the binomial theorem, as $1 \pm \frac{1}{2}\sqrt{(2/n)}$; hence

$$\mu_c - \mu = \pm 0{\cdot}707\frac{\mu}{\sqrt{n}} \quad \text{(S.E.)}. \tag{10}$$

Similarly, from (9),

$$r_c - r = \pm\tfrac{1}{2}\alpha r\sqrt{\frac{2}{n}} = \pm 0{\cdot}4769\frac{r}{\sqrt{n}} \quad \text{(P.E.)}. \tag{11}$$

These results may be summarized as follows:

the standard error of the computed quantity, μ_c, is $0{\cdot}707\mu/\sqrt{n}$;

the probable error of the computed quantity, r_c, is $0{\cdot}4769r/\sqrt{n}$.

(iii) *Peters's formula.* Let η be the average error of the N measures; the relation between η and μ is

$$\mu = \eta\sqrt{(\tfrac{1}{2}\pi)}.$$

In Peters's formula (3) let η_c denote the calculated value for the sample given by

$$\eta_c = \frac{\Sigma\,|\,v_i\,|}{\sqrt{\{n(n-1)\}}}.$$

Let e denote the error in μ_c calculated according to (3); here e is not to be confused with the 'e' in (ii). Then

$$e \equiv \mu_c - \mu = \eta_c \sqrt{(\tfrac{1}{2}\pi)} - \mu, \qquad (12)$$

from which

$$e^2 + 2\mu e = \frac{\pi}{2} \frac{\{\Sigma \mid v_i \mid\}^2}{n(n-1)} - \mu^2$$

$$= \frac{\pi}{2n(n-1)} [\Sigma v_i^2 + \Sigma \mid v_i \mid \Sigma \mid v_j \mid] - \mu^2 \quad (j \neq i).$$

With the notation of (ii),

$$M(e^2) + 2\mu M(e) = \frac{\pi}{2n(n-1)} [\Sigma M(v_i^2) + \Sigma M(\mid v_i \mid)\, M(\mid v_j \mid)] - \mu^2.$$

Let μ_0 be the standard error of the e's; again, not to be confused with the standard error in (ii).

As in (ii), $$M(e) = 0 \quad \text{and} \quad M(v_i^2) = \frac{1}{2H^2}.$$

Also $$M(\mid v \mid) = \frac{2H}{\sqrt{\pi}} \int_0^\infty v\, e^{-H^2 v^2}\, dv = \frac{1}{H\sqrt{\pi}}.$$

Hence

$$\mu_0^2 = \frac{\pi}{2n(n-1)} \left[n\left(\frac{1}{2H^2}\right) + n\left(\frac{1}{H\sqrt{\pi}}\right)(n-1)\left(\frac{1}{H\sqrt{\pi}}\right) \right] - \mu^2,$$

or, by (4), $$\mu_0^2 = \frac{(\pi - 2)\mu^2}{2n} \equiv \left\{0{\cdot}7555 \frac{\mu}{\sqrt{n}}\right\}^2. \qquad (13)$$

From (12) and (13) we have

$$e \equiv \mu_c - \mu = \pm 0{\cdot}7555 \frac{\mu}{\sqrt{n}} \quad \text{(S.E.)}, \qquad (14)$$

the expression on the right being the standard error associated with the e's.

Similarly, as in (ii), the corresponding formula in terms of probable error is

$$r_c - r = \pm 0{\cdot}5096 \frac{r}{\sqrt{n}} \quad \text{(P.E.)}. \qquad (15)$$

(iv) *Comparison of formulae* (2) *and* (3). Considering the two formulae (10) and (14) relating to the two methods of calculating the standard error by means of (2) and (3), we can write (10) and (14) respectively as

$$\frac{\text{S.E. of } \mu_c}{\mu} = \frac{0{\cdot}707}{\sqrt{n}}$$

and $$\frac{\text{S.E. of } \mu_c}{\mu} = \frac{0{\cdot}7555}{\sqrt{n}}.$$

These formulae show that, in general, formula (2) is slightly superior to formula (3); the latter, however, as we have seen, has the advantage as regards simplicity of calculation.

Similarly, from (11) and (15) respectively,

$$\frac{\text{P.E. of } r_c}{r} = \pm\frac{0{\cdot}4769}{\sqrt{n}} \tag{16}$$

and

$$\frac{\text{P.E. of } r_c}{r} = \pm\frac{0{\cdot}5096}{\sqrt{n}}, \tag{17}$$

which show that (2) is slightly superior to (3).

The preceding formulae emphasize the approximate nature of the calculated standard and probable errors when the number of measures for a sample is small. Consider, for example, (16) or (17) when $n=9$; from either formula the probable error of r_c is about 16 or 17 % of the true probable error. Also, since the P.E., r_a, of the arithmetic mean is $r_c/\sqrt{n}$, these results apply equally to the computed value of the arithmetic mean. For larger values of n, say $n=100$, the calculated percentage is about 5 % and, in this case, the values of the calculated probable errors may be taken to represent with sufficient accuracy the true values of the probable errors.

CHAPTER 5

MEASURES OF PRECISION FOR WEIGHTED OBSERVATIONS

5·01. Weighting of observations

The weighting of observations is a process which takes into account the superior quality of one set of observations over another set of observations, the latter, made perhaps under poor conditions or perhaps by a relatively inexperienced observer; even if the external conditions do not vary from one set to another, the superiority of one set may consist in the larger number of individual measures as compared with the number of individual measures of a second set. Our problem is to investigate methods of combining different sets of observations of varying reliability, and this is achieved in terms of *weights* applied to the final results of each set of measures.

The methods of weighting which will be considered in turn are as follows: (i) weighting according to the number of individual observations forming a set; (ii) arbitrary weighting when the external conditions vary from one set to another; (iii) weighting according to the standard errors or probable errors derived for the results of the several sets.

5·02. Weighting according to the number of observations in a set

We consider, for example, several sets of measures of the position angle of a double star made by the same observer on nights when the observing conditions are assumed to be equally favourable (see § 4·12); all the individual measures are then to be regarded as equally trustworthy.

On the first night the observer makes n_1 single measures and, all these being deemed to be equally reliable, he takes the arithmetic mean, a_1, to represent the value of the position angle resulting from the night's observations; then, if S_1 denotes the sum of the individual measures, $n_1 a_1 = S_1$. Generally, for the i-th night, the corresponding quantities are n_i, a_i and S_i with the relation

$$S_i = n_i a_i. \tag{1}$$

Suppose that he observes on six nights; the total number of individual measures is $n_1 + n_2 + \ldots + n_6 \equiv \Sigma n_i$, and the sum of the individual measures is $S_1+S_2+\ldots+S_6\equiv\Sigma S_i$. Hence the arithmetic mean, a, of all the single measures is given by

$$a=\Sigma S_i/\Sigma n_i, \tag{2}$$

or, by means of (1), $$a=\Sigma n_i a_i/\Sigma n_i. \tag{3}$$

Instead of calculating a, according to (2), by adding all the individual measures for the six nights and then dividing by the total number of individual measures, it is much more convenient for the observer to derive the values of a_i resulting from each night's observations and then to calculate a by means of (3). In this case the numbers $n_1, n_2, \ldots$ are the *weights* of the quantities $a_1, a_2, \ldots$ and are commonly denoted by $w_1, w_2, \ldots$; the total weight, W, is $n_1+n_2+\ldots$ or $w_1+w_2+\ldots$, so that

$$W=\Sigma w_i. \tag{4}$$

Then (3) becomes $$a=\frac{\Sigma w_i a_i}{W}; \tag{5}$$

a is called the *weighted mean* of the several results $a_1, a_2, \ldots$ obtained from the different series of observations. It will be shown later that the weighted mean, a, is the most probable value of the unknown.

As has been stated, the individual measures are assumed to be equally trustworthy. Now, if the total number of individual measures is denoted by N ($\equiv\Sigma n_i$) and $\alpha_1, \alpha_2, \ldots, \alpha_N$ are the individual measures, then, by (2),

$$a=\frac{\alpha_1+\alpha_2+\ldots+\alpha_N}{N}, \tag{6}$$

which shows that a is the weighted mean of the N measures if the weight to be attached to each measure is unity. Thus, we can describe the equally trustworthy individual measures as *measures of unit weight*.

Let μ_0 and h_0 denote the standard error and the corresponding modulus of the individual measures (each of unit weight); then, by 4·04 (3),

$$\mu_0^2=\frac{1}{2h_0^2}. \tag{7}$$

Consider now the arithmetic mean, a_i, of the i-th set of measures, w_i in number; then, by 4·06 (3), the standard error of a_i is given by

$$\mu_i^2=\frac{\mu_0^2}{w_i}=\frac{1}{w_i(2h_0^2)}, \tag{8}$$

and, if h_i is the modulus associated with μ_i,

$$\mu_i^2=\frac{1}{2h_i^2}. \tag{9}$$

From (8) and (9) we have $h_i^2 = w_i h_0^2$. (10)

Thus, h_i is given in terms of the known value, w_i, and of h_0 which is regarded as a constant in the circumstances concerned.

These formulae will be used later.

Returning to (5) we remark that, in calculating the arithmetic mean, a, only the *ratios* of the weights need be considered, for if w_1, for example, is the smallest weight, we can write (5) in the convenient form

$$a = \frac{a_1 + a_2\left(\frac{w_2}{w_1}\right) + a_3\left(\frac{w_3}{w_1}\right) + \dots}{1 + \frac{w_2}{w_1} + \frac{w_3}{w_1} + \dots}. \quad (11)$$

Thus, a is obtained in terms of the *relative weights*, the weight of a_1 being unity.

5·03. Arbitrary weighting

We continue to consider the measures of the position angle of a double star, but now we suppose that the observing conditions vary from night to night.

First, we suppose that the same number of single measures are made on each night and that the arithmetic means for the several nights are $a_1, a_2, \dots$. If a_1, for example, refers to a night when the observing conditions are good and a_2 when the observing conditions are poor, then on general grounds a_1 would be accorded a greater degree of reliability than a_2. The observing conditions are, of course, beyond the control of the observer and the best he can do—unless he adopts the system of weighting, (iii), mentioned in § 5·01 and illustrated in § 5·10 (ii) and (iii)—is to assess the effects of these conditions on his measures according to an arbitrary scale, the 'scale of seeing', as it is called. In the present instance we may suppose that the scale is taken to be: 1, 2, 3, 4, 5 or, symbolically, s_i; on this scale s_1 denotes poor conditions and s_5 denotes excellent conditions, with assessments s_2, s_3 and s_4 for intermediate conditions. As a result of experience the observer then regards a particular measure when the 'seeing' is s_4, for example, as equivalent to 4 measures made on a night when the 'seeing' is s_1; thus, the weight in this case is 4. In general, the weighted mean, a, is given by

$$a = \frac{\Sigma s_i a_i}{\Sigma s_i},$$

or, by 5·02 (5),

$$a = \frac{\Sigma w_i a_i}{W},$$

in which $w_i = s_i$ and $W = \Sigma w_i$.

Secondly, if the number of individual observations varies from night to night, n_i corresponding to a_i with the 'seeing' denoted by s_i, the weight to be attached to a_i is $n_i s_i$; then a is given by

$$a = \Sigma n_i s_i a_i / \Sigma n_i s_i$$

or

$$a = \frac{\Sigma w_i a_i}{\Sigma w_i},$$

where $w_i = n_i s_i$.

In both cases we can regard the measures made when the seeing is s_1 as measures of unit weight, with modulus h_0 and standard error μ_0. Then, as in the previous section, if h_i and μ_i refer to an arithmetic mean, a_i, with weight w_i, we have

$$\mu_i^2 = \frac{\mu_0^2}{w_i} \tag{1}$$

and

$$h_i^2 = w_i h_0^2. \tag{2}$$

5·04. Weighting according to standard errors or probable errors

As an example consider the determination of the most probable value of the velocity of light (denoted here by a) derived from measures made by n different observers using the same or different experimental methods.

An observer, A_1, makes a series of equally reliable measures from which he finds, by the rules given in § 4·11, (i) the most probable value of the unknown, namely, the arithmetic mean, a_1, of his measures, and (ii) the standard error, μ_1, or the probable error, r_1, of the arithmetic mean, a_1. It is to *be particularly noted that here μ_1 denotes the value of μ_a associated with a_1*. Thus, if the individual measures, p in number, are $x_1, x_2, \ldots, x_p$, then

$$a_1 = \frac{\Sigma x_i}{p};$$

also, if v_i $(\equiv x_i - a_1)$ is the residual corresponding to x_i, then, by 4·06 (4);

$$\mu_1^2 \equiv \mu_a^2 = \frac{[vv]}{p(p-1)}.$$

The probable error, $r_1 \equiv r_a$, of the arithmetic mean, a_1, is given by

$$r_1 = 0{\cdot}6745 \mu_1.$$

The values of μ_1 and r_1 may, of course, be found by means of Peters's formula.

The definitive results of this observer are:

(i) velocity of light $=a_1$,

(ii) standard error of $a_1=\mu_1$, or probable error of $a_1=r_1$.

Considering now the work of the n observers we have the *individual results* represented by the arithmetic means

$$a_1, a_2, \ldots, a_n,$$

together with the corresponding standard errors

$$\mu_1, \mu_2, \ldots, \mu_n,$$

or probable errors $\quad r_1, r_2, \ldots, r_n,$

and the associated moduli

$$h_1, h_2, \ldots, h_n.$$

In general, the several values of μ_i $(i=1, 2, \ldots, n)$ will all be different; in other words, the arithmetic means, a_i, will be determined with varying degrees of precision represented by the weights, w_i, to be attached to the values of the arithmetic means. Thus, we can suppose that a result, a_i, is equivalent to w_i measures of unit weight. Further, if μ_0 and h_0 refer to a particular determination of a_i, regarded as of unit weight, then by 5·03 (1) and (2), h_i being the modulus corresponding to μ_i,

$$w_i=\frac{\mu_0^2}{\mu_i^2}=\frac{h_i^2}{h_0^2}. \tag{1}$$

The weighted mean, a, of the n results is given by

$$a\equiv\frac{\Sigma w_i a_i}{W}=\frac{\Sigma a_i\left(\frac{\mu_0}{\mu_i}\right)^2}{\Sigma\left(\frac{\mu_0}{\mu_i}\right)^2}.$$

Suppose now that the *largest* of the standard errors of the arithmetic means is μ_1; then, discarding the factor μ_0^2 common to numerator and denominator of the above expression for a, we have

$$a=\frac{\frac{a_1}{\mu_1^2}+\frac{a_2}{\mu_2^2}+\ldots+\frac{a_n}{\mu_n^2}}{\frac{1}{\mu_1^2}+\frac{1}{\mu_2^2}+\ldots+\frac{1}{\mu_n^2}}$$

$$=\frac{a_1+a_2\left(\frac{\mu_1}{\mu_2}\right)^2+\ldots+a_n\left(\frac{\mu_1}{\mu_n}\right)^2}{1+\left(\frac{\mu_1}{\mu_2}\right)^2+\ldots+\left(\frac{\mu_1}{\mu_n}\right)^2}. \tag{2}$$

Here $(\mu_1/\mu_i)^2$ is the *relative weight* of the arithmetic mean, a_i, the weight of a_1 being unity.

Since
$$\mu_1/\mu_i = r_1/r_i,$$
which follows from the relation $r_a = 0{\cdot}6745\mu_a$, the formulae (1) and (2) are, in terms of probable error,
$$w_i = \frac{r_0^2}{r_i^2} = \frac{h_i^2}{h_0^2} \tag{3}$$
and
$$a = \frac{a_1 + a_2\left(\frac{r_1}{r_2}\right)^2 + \ldots + a_n\left(\frac{r_1}{r_n}\right)^2}{1 + \left(\frac{r_1}{r_2}\right)^2 + \ldots + \left(\frac{r_1}{r_n}\right)^2}. \tag{4}$$
The relative weight of the arithmetic mean, a_i, is $(r_1/r_i)^2$.

5·05. The weighted mean is the most probable value of the unknown

This theorem can be proved very simply as follows. In 5·02 (6), a is the arithmetic mean of *all* the individual measures, N in number and all of equal weight, made by the n observers and is accordingly the most probable value of the unknown. But, also, by 5·02 (3), $a = \Sigma n_i a_i / \Sigma n_i$ or, writing w for n,
$$a = \frac{\Sigma w_i a_i}{\Sigma w_i}.$$
Hence the weighted mean is the most probable value of the unknown.

An alternative proof is as follows. Consider the n separate and independent determinations, a_i, of a quantity whose true value is x, a_i being the arithmetic mean of the i-th set of individual measures, p_i in number.

We have n separate equations, each with its appropriate weight, of the form
$$a_i - x = \epsilon_i \quad \text{(weight, } w_i\text{)},$$
where ϵ_i is the error of a_i. The standard error associated with a_i is μ_i, with h_i as the corresponding modulus; then, by 5·04 (1),
$$h_i^2 = w_i h_0^2, \tag{1}$$
in which h_0 refers to a particular determination of a_i of unit weight.

If k is a small quantity related to the degree of accuracy attainable by the measuring apparatus, the probability that the error of a_i lies between $\epsilon_i - \frac{1}{2}k$ and $\epsilon_i + \frac{1}{2}k$ is
$$\frac{kh_i}{\sqrt{\pi}} \exp\{-h_i^2 \epsilon_i^2\} \quad \text{or} \quad \frac{kh_i}{\sqrt{\pi}} \exp\{-h_i^2 (a_i - x)^2\}.$$

Hence the probability, P, that the n errors $\epsilon_1, \ldots, \epsilon_n$ occur is given by

$$P=\left(\frac{k}{\sqrt{\pi}}\right)^n \Pi(h_i)\exp\{-\Sigma h_i^2(a_i-x)^2\}.$$

The most probable value of the unknown is that which makes P a maximum or that which makes $E\equiv\Sigma h_i^2(a_i-x)^2$ a minimum. The conditions are (i) $dE/dx=0$ and (ii) $d^2E/dx^2>0$. The second condition is obviously satisfied; the first condition gives $\Sigma h_i^2(a_i-x)=0$, whence the most probable value of x is $\Sigma h_i^2 a_i/\Sigma h_i^2$, or, by means of (1), $\Sigma w_i a_i/\Sigma w_i$. Accordingly, the most probable value of the unknown is the weighted mean, a.

5·06. Reduction of weighted equations to equations of unit weight

We write a typical equation of weight w as

$$X-x=\epsilon, \tag{1}$$

where X is one of the values $a_1, a_2, \ldots, a_n$, each a being derived from a set of observations, and ϵ is the corresponding error. Let μ and h be, respectively, the standard error and modulus of precision associated with ϵ so that

$$\mu^2=1/(2h^2). \tag{2}$$

Now, in general, the probability, p, that an error occurs between ϵ and $\epsilon+d\epsilon$ is given by

$$p=\frac{h}{\sqrt{\pi}}e^{-h^2\epsilon^2}\,d\epsilon. \tag{3}$$

Define an error, ζ, by $\xi=c\epsilon$ $(c>1)$; then we shall assume that ξ is associated with an equation of unit weight, with modulus h_0 and standard error μ_0. In terms of ξ, (3) becomes

$$p=\frac{h}{c\sqrt{\pi}}\exp\left\{-\frac{h^2}{c^2}\xi^2\right\}d\xi;$$

thus p is the probability that an error, ξ, occurs between ξ and $\xi+d\xi$, the modulus being h/c. But the modulus is also h_0; hence

$$h_0=\frac{h}{c},$$

which becomes, by means of (2) and the corresponding formula for μ_0, namely, $\mu_0^2=1/(2h_0^2)$,

$$c=\frac{h}{h_0}=\frac{\mu_0}{\mu}.$$

Now, the weight, w, is defined by

$$w=\frac{h^2}{h_0^2}=\frac{\mu_0^2}{\mu^2}.$$

Hence

$$c=\sqrt{w}. \tag{4}$$

Multiplying (1) by c; then, by means of (4),

$$\sqrt{w}\,(X-x)=\sqrt{w}\,\epsilon\equiv\xi,$$

which is an equation of unit weight. Thus, in general, to reduce the equation

$$x_i-x=\epsilon_i \quad \text{(weight, } w_i\text{)}$$

to an equation of unit weight, we multiply throughout by $\sqrt{w_i}$ and so obtain the equation

$$\sqrt{w_i}\,(x_i-x)=\sqrt{w_i}\,\epsilon_i. \tag{5}$$

Note. An observational equation must not be multiplied by any factor other than the weight when it is to be combined with other equations of a like nature, for, if (5) is multiplied throughout by 2, for example, it becomes an equation with fictitious weight $4w_i$.

5·07. The formula $\mu_0^2=\dfrac{[wvv]}{n-1}$

We begin with the n equations of the form

$$a_i-x=\epsilon_i \quad \text{(weight, } w_i\text{)},$$

where, as before, a_i represents the result from a particular set of measures. Reduced to unit weight this equation becomes

$$\sqrt{w_i}\,(a_i-x)=\sqrt{w_i}\,\epsilon_i\equiv\xi_i.$$

As we have seen the most probable value of the unknown is the weighted mean, a. Let v_i be the residual corresponding to the result a_i; then

$$a_i-a=v_i$$

and

$$\sqrt{w_i}\,(a_i-a)=\sqrt{w_i}\,v_i\equiv V_i. \tag{1}$$

Thus, the residual V_i corresponds to the error ξ_i.

There are n equations of the form (1), each of unit weight.

Now, μ_0 is the standard error associated with the errors ξ_i, the corresponding residuals being V_i; hence, by 4·05 (5), which relates to equations of equal weight,

$$\mu_0^2=\frac{\Sigma V_i^2}{n-1},$$

or, by (1),
$$\mu_0^2 = \frac{\Sigma w_i v_i^2}{n-1},$$

or, in the Gaussian notation,

$$\mu_0^2 = \frac{[wvv]}{n-1}. \tag{2}$$

This is an important formula which enables us to calculate the standard error of unit weight from the residuals and weights of the n equations.

Returning to (1) we can obtain, by a similar process, Peters's formula for μ_0. Now, by 4·08 (5),

$$\mu_0 = 1{\cdot}2533 \frac{\Sigma |V_i|}{\sqrt{\{n(n-1)\}}},$$

or, from (1),
$$\mu_0 = 1{\cdot}2533 \frac{[\sqrt{w}\, v]}{\sqrt{\{n(n-1)\}}}, \tag{3}$$

in which the numerical values of the residuals, v, are to be used, irrespective of sign.

The equations (2) and (3) provide alternative methods for calculating the standard error of unit weight.

Since in general $r = 0{\cdot}6745\mu$, the formulae (2) and (3) can be expressed in terms of the probable error, r_0, of unit weight. Thus

$$r_0 = 0{\cdot}6745 \sqrt{\left\{\frac{[wvv]}{n-1}\right\}} \tag{4}$$

and
$$r_0 = 0{\cdot}8453 \frac{[\sqrt{w}\, v]}{\sqrt{\{n(n-1)\}}}. \tag{5}$$

As mentioned in §5·04, it is convenient in practice to associate μ_0 and r_0 with the equation of lowest weight.

5·08. The precision of the weighted mean

The weighted mean, a, is given by

$$a = \left(\frac{w_1}{W}\right) a_1 + \left(\frac{w_2}{W}\right) a_2 + \ldots + \left(\frac{w_n}{W}\right) a_n, \tag{1}$$

in which a_1, a_2, ... are the results of the n independent series of measures; μ_1, μ_2, ... are the standard errors associated with the results a_1, a_2, ... As before, μ_0 is the standard error arising from presumed measures of unit weight.

In observations such as those described in §§5·02 and 5·03 the weights w_i are known; the standard errors μ_i associated with the results a_i are then found from the formula

$$w_i \mu_i^2 = \mu_0^2 \tag{2}$$

in terms of μ_0. In the type of investigation described in §5·04 each observer records the particular standard error μ_i and then the corresponding weight w_i to be attached to the corresponding result can be found by means of (2) in terms of μ_0.

Let μ_a be the standard error of the weighted mean given by (1). Then, by the theorem of §3·07, the n results $a_1, a_2, \ldots$ being independent, we have

$$\mu_a^2 = \left(\frac{w_1}{W}\right)^2 \mu_1^2 + \left(\frac{w_2}{W}\right)^2 \mu_2^2 + \ldots + \left(\frac{w_n}{W}\right)^2 \mu_n^2 = \Sigma \left(\frac{w_i}{W}\right)^2 \mu_i^2.$$

This formula becomes, by means of (2),

$$\mu_a^2 = \frac{\mu_0^2}{W^2} \Sigma w_i = \frac{\mu_0^2}{W}.$$

Hence

$$\mu_a = \frac{\mu_0}{\sqrt{W}}, \tag{3}$$

in which μ_0 is given by (2) or (3) of the previous section.

The probable error, r_a, of the weighted mean is given by

$$r_a = \frac{r_0}{\sqrt{W}},$$

in which r_0 is given by (4) or (5) of the previous section.

5·09. Summary of formulae

The results from n independent series of measures are $a_1, a_2, \ldots, a_n$; the corresponding weights are $w_1, w_2, \ldots, w_n$ which are found, according to circumstances, by the considerations described in §§5·02–5·04.

The weighted mean, a, is given by

$$a = \frac{w_1 a_1 + w_2 a_2 + \ldots + w_n a_n}{W},$$

where $W = w_1 + w_2 + \ldots + w_n$.

The residuals $v_1 \equiv a_1 - a$, $v_2 \equiv a_2 - a$, ... are obtained.

As in §4·11, it is convenient to summarize the various precision formulae, first, in terms of the sum of the squares of the residuals and, secondly, in terms of the sum of the numerical values of the residuals.

Method I:

$$\mu_0 = \sqrt{\left\{\frac{[wvv]}{n-1}\right\}},$$

$$r_0 = 0{\cdot}6745\sqrt{\left\{\frac{[wvv]}{n-1}\right\}},$$

$$\mu_a = \sqrt{\left\{\frac{[wvv]}{(n-1)W}\right\}},$$

$$r_a = 0{\cdot}6745\sqrt{\left\{\frac{[wvv]}{(n-1)W}\right\}}.$$

Method II:

$$\mu_0 = 1{\cdot}2533\,\frac{[\sqrt{w}\,v]}{\sqrt{\{n(n-1)\}}},$$

$$r_0 = 0{\cdot}8453\,\frac{[\sqrt{w}\,v]}{\sqrt{\{n(n-1)\}}},$$

$$\mu_a = 1{\cdot}2533\,\frac{[\sqrt{w}\,v]}{\sqrt{\{n(n-1)\,W\}}},$$

$$r_a = 0{\cdot}8453\,\frac{[\sqrt{w}\,v]}{\sqrt{\{n(n-1)\,W\}}}.$$

5·10. Examples

(i) *Weighting by number of measures.* We consider the seven measures of the position angle, θ, of the double star β 1077 given in Table 8 (p. 83), but we now suppose that the number of individual measures on each night are: 12, 12, 8, 16, 8, 12, 12. Since the ratios of the weights only are required, we can discard the common factor, 4; the resulting weights, w, are shown in the second column of Table 10. As in § 4·12, we write

$$\theta = 300° + x°.$$

Table 10. *Measures of position angle*

			v		wv			
x	w	wx	+	−	+	−	wv^2	$\sqrt{w}v$
1·9	3	5·7	—	6·1	—	18·3	111·6	10·6
2·2	3	6·6	—	5·8	—	17·4	100·8	10·0
11·8	2	23·6	3·8	—	7·6	—	28·8	5·4
9·5	4	38·0	1·5	—	6·0	—	9·0	3·0
9·9	2	19·8	1·9	—	3·8	—	7·2	2·7
12·9	3	38·7	4·9	—	14·7	—	72·0	8·5
9·5	3	28·5	1·5	—	4·5	—	6·7	2·6
	20	**160·9**	**13·6**	**11·9**	**36·6**	**35·7**	**336·1**	**42·8**

Summary: $n=7$; $W \equiv \Sigma w = 20$; $\Sigma wx = 160{\cdot}9$; $[wvv] = 336{\cdot}1$; $[\sqrt{w}\,|v|] = 42{\cdot}8$.

The weighted mean is given by

$$a=\frac{\Sigma wx}{W}=\frac{160{\cdot}9}{20}=8{\cdot}0 \quad \text{(to one decimal place).}$$

The residuals according to sign are found in the fourth column. The near equality (numerical) of the positive and negative sums in columns 6, 7 affords a check on the work up to this point; the small difference is due to taking a to one decimal place only.

The entries in the final column are the square roots of the corresponding entries in the penultimate column and are written down by means of Barlow's tables.

We apply the rules as summarized in § 5·09:

	Method I	Method II
μ_0	$\sqrt{\frac{336{\cdot}1}{6}}=7{\cdot}48$	$\frac{1{\cdot}2533\times 42{\cdot}8}{\sqrt{42}}=8{\cdot}27$
μ_a	$\frac{\mu_0}{\sqrt{20}}=1{\cdot}67$	$\frac{\mu_0}{\sqrt{20}}=1{\cdot}85$
r_0	$0{\cdot}6745\mu_0=5{\cdot}05$	$0{\cdot}6745\mu_0=5{\cdot}58$
r_a	$\frac{r_0}{\sqrt{20}}=1{\cdot}13$	$\frac{r_0}{\sqrt{20}}=1{\cdot}25$

The results by Method I are, to one decimal place:

$$\theta=308^\circ{\cdot}0\pm 1^\circ{\cdot}7 \quad \text{(S.E.)}$$

and

$$\theta=308^\circ{\cdot}0\pm 1^\circ{\cdot}1 \quad \text{(P.E.).}$$

The values of the standard error and probable error given by Method II are a little larger than the values found by Method I.

(ii) *Weighting according to probable error* (*first example*). We consider the eight separate and independent determinations of the velocity of light, made between 1951 and 1954, in Essen's list.† We write the velocity of light, c, in km./sec., as

$$c=299{,}790+x.$$

The values of x are in the first column of Table 11 and the published probable errors are in the second column.

The probable errors vary between $\pm 0{\cdot}2$ and $\pm 7{\cdot}0$; it is convenient to take an intermediate value, $r_0=\pm 3{\cdot}0$, as of unit weight. The weight of the first determination is then $(3{\cdot}0/0{\cdot}2)^2$ or 225; the remaining weights are in the third column.

† L. Essen, *Endeavour*, **15** (1956), 90.

Table 11. *Measures of the velocity of light*

x	P.E. $\pm$	w	wx	v	wv	wv^2	$\sqrt{w}\,\vert v\vert$
+ 3·1	0·2	225·0	+697·5	+ 0·1	+22·5	2·25	1·50
+ 4·2	1·9	2·49	+ 10·5	+ 1·2	+ 3·0	3·60	1·90
+ 2·6	0·7	18·37	+ 47·8	− 0·4	− 7·3	2·92	1·71
−14·0	7·0	0·18	− 2·5	−17·0	− 3·1	52·70	7·26
+ 3·0	0·3	100·0	+300·0	0·0	0·0	0·0	0·0
− 0·2	3·0	1·0	− 0·2	− 3·2	− 3·2	10·24	3·20
+ 5·0	3·1	0·94	+ 4·7	+ 2·0	+ 1·9	3·80	1·95
+ 2·0	6·0	0·25	+ 0·5	− 1·0	− 0·2	0·20	0·45
		348·2	**1058·3**			**75·71**	**17·97**

Summary:

$n=8$; $W\equiv\Sigma w=348{\cdot}2$; $\Sigma wx=1058{\cdot}3$; $[wvv]=75{\cdot}71$; $[\sqrt{w}\,|v|]=17{\cdot}97$.

The weighted mean $a\equiv\dfrac{\Sigma wx}{W}=\dfrac{1058{\cdot}3}{348{\cdot}2}=3{\cdot}0$; the most probable value of the velocity of light is then 299,790 + 3 or 299,793.

We calculate μ_a and r_a:

	Method I	Method II
μ_a	$\sqrt{\left\{\dfrac{75{\cdot}71}{7\times 348{\cdot}2}\right\}}=0{\cdot}18$	$\dfrac{1{\cdot}2533\times 17{\cdot}97}{\sqrt{\{7\times 8\times 348{\cdot}2\}}}=0{\cdot}16$
r_a	$0{\cdot}6745\mu_a \quad =0{\cdot}12$	$0{\cdot}11$

Taking the larger values of μ_a and r_a we obtain the following results:

$$c=299{,}793{\cdot}0\pm 0{\cdot}18 \quad \text{(S.E.)},$$

$$c=299{,}793{\cdot}0\pm 0{\cdot}12 \quad \text{(P.E.)}.$$

It will be observed that the first and fifth measures (with relative weights 225 and 100) make by far the biggest contribution to the results.

(iii) *Weighting according to probable error* (*second example*). As a second example we take the first eight determinations of the solar parallax, Π, derived from declination observations† and tabulated by Spencer Jones.

We write Π (in seconds of arc) as

$$\Pi=8{\cdot}7900+10^{-4}x.$$

The values of x and the probable errors of the corresponding determinations are given in Table 12. No. 4 has the largest probable

† H. Spencer Jones, *Mon. Not. R. Astr. Soc.* **101** (1941), 360; fourteen determinations are given, but for purposes of illustration we take the first eight of the tabular results only.

error and we take this result to be of unit weight. The weight, w, of no. 1 is then $(\frac{67}{26})^2$ or 6·64; the other weights are found in the same way.

The weighted mean is: $a = \dfrac{223 \cdot 5}{34 \cdot 00} = 6 \cdot 6$. The residuals are then entered in the table.

Table 12. *Parallax measures*

No.	x	P.E. $\pm$	w	wx +	wx −	v	wv +	wv −	wv^2	$\sqrt{w}v$
1	+ 34	26	6·64	225·8	—	+ 27·4	181·9	—	4,984	70·6
2	+ 35	40	2·80	98·0	—	+ 28·4	79·5	—	2,258	47·6
3	+ 8	26	6·64	53·1	—	+ 1·4	9·3	—	13	3·6
4	+123	67	1·00	123·0	—	+116·4	116·4	—	13,549	116·4
5	+ 6	33	4·12	24·7	—	− 0·6	—	2·5	1	1·0
6	− 26	39	2·95	—	76·7	− 32·6	—	96·2	3,136	56·0
7	− 14	28	5·73	—	80·2	− 20·6	—	118·0	2,431	49·3
8	− 35	33	4·12	—	144·2	− 41·6	—	171·4	7,130	84·4
			34·00	**524·6**	**301·1**		**387·1**	**388·1**	**33,502**	**428·9**

Summary: $n=8$; $W \equiv \Sigma w = 34 \cdot 0$; $\Sigma wx = +223 \cdot 5$; $[wvv] = 33{,}502$; $[\sqrt{w}\,|v|] = 428 \cdot 9$.

To four places of decimals, $\Pi = 8'' \cdot 7900 + 0'' \cdot 0007 = 8'' \cdot 7907$.
We calculate μ_a and r_a:

	Method I	Method II
μ_a	$\sqrt{\left\{\dfrac{33502}{7 \times 34}\right\}} = 11 \cdot 9$	$\dfrac{1 \cdot 2533 \times 428 \cdot 9}{\sqrt{(56 \times 34)}} = 12 \cdot 3$
r_a	$0 \cdot 6745 \mu_a = 8 \cdot 0$	$8 \cdot 3$

The unit in which μ_a and r_a are expressed is $0'' \cdot 0001$.
In terms of probable error the evaluation of Π is given by

$$\Pi = 8'' \cdot 7907 \pm 0'' \cdot 0008 \quad \text{(P.E.)}.$$

CHAPTER 6

EQUATIONS OF CONDITION IN SEVERAL UNKNOWNS

6·01. Linear equations of condition

In many problems the measured quantity is related to several unknowns, frequently by means of a *linear equation*. For example, the determination of the parallax, Π, of a star from photographic observations is dependent on the solution of a set of linear equations in which Π and the component, μ_α, of the star's proper motion are unknowns, each equation being of the form†

$$a\Pi + b\mu_\alpha = m.$$

Here m is a combination of a measured quantity and a known constant, c, the latter being determined by means of an independent procedure; it is convenient to refer to m simply as the *measured quantity*; also, corresponding to a particular value of m, a and b have known numerical values. In § 6·20 the solution of a group of equations with the above form will be considered in greater detail.

In dealing with principles we shall consider n equations in three unknowns (x, y and z) of the form

$$a_i x + b_i y + c_i z = m_i, \tag{1}$$

in which m_i is the measured quantity and the corresponding values a_i, b_i and c_i are known. It is assumed that n is greater than the number of unknowns; in this case $n > 3$.

We refer to (1) as an *equation of condition*.

Our problem is to determine the most probable values of the unknowns x, y and z together with an assessment of their precision.

6·02. Equations of condition in functional form

In some problems the equation of condition is of functional form. Denote the unknowns in this case by X, Y and Z and the measured quantity by l; then, the equation of condition for the i-th measure is

$$\phi_i(X, Y, Z) = l_i, \tag{1}$$

† W. M. Smart, *Spherical Astronomy*, 4th ed. (Cambridge University Press, p. 311 (formula (89)).

in which the function ϕ_i contains one or more constants, as in 6·01 (1), associated with l_i.

In practice we can make progress as a rule only if (1) can be converted into a linear form, and this involves deriving, as best we can, approximate values of X, Y and Z which we denote by X_0, Y_0 and Z_0. Write

$$X=X_0+x,\quad Y=Y_0+y,\quad Z=Z_0+z,$$

where x, y and z are assumed to be small quantities whose squares and products can be neglected. By Taylor's theorem, (1) becomes

$$\phi_i(X,Y,Z)=\phi_i(X_0,Y_0,Z_0)+x\left(\frac{\partial\phi_i}{\partial X}\right)_0+y\left(\frac{\partial\phi_i}{\partial Y}\right)_0+z\left(\frac{\partial\phi_i}{\partial Z}\right)_0=l_i. \quad (2)$$

The function ϕ_i being known, the values of $(\partial\phi_i/\partial X)_0$, etc., can be calculated; denote these values by a_i, b_i and c_i and write

$$m_i=l_i-\phi_i(X_0,Y_0,Z_0).$$

Then (2) becomes

$$a_ix+b_iy+c_iz=m_i, \quad (3)$$

which is the equation of condition in linear form.

To illustrate the reduction of an equation of condition in functional form to the linear form consider the Hartmann-Cornu formula† which, in terms of the usual spectroscopic notation, is

$$\phi_i(m_0,c,\lambda_0;\lambda_i)\equiv m_0+\frac{c}{(\lambda_i-\lambda_0)^p}=l_i; \quad (4)$$

here m_0, c and λ_0 are 'plate constants', p is an instrumental constant which is supposed known (for many instruments its value is close to unity) and l_i is a measured quantity corresponding to a known wavelength λ_i. The plate constants are the unknowns; write $m_0=X$, $c=Y$ and $\lambda_0=Z$; then (4) becomes

$$\phi_i(X,Y,Z;\lambda_i)\equiv X+\frac{Y}{(\lambda_i-Z)^p}=l_i. \quad (5)$$

In the simple case when $p=1$, the approximate values of X, Y and Z are readily obtained from three equations with widely separated values of λ_i; thus

$$X+\frac{Y}{\lambda_1-Z}=l_1,\quad X+\frac{Y}{\lambda_2-Z}=l_2,\quad X+\frac{Y}{\lambda_3-Z}=l_3, \quad (6)$$

from which

$$l_1-l_2=\frac{Y(\lambda_2-\lambda_1)}{(\lambda_1-Z)(\lambda_2-Z)}\quad\text{and}\quad l_2-l_3=\frac{Y(\lambda_3-\lambda_2)}{(\lambda_2-Z)(\lambda_3-Z)}; \quad (7)$$

† F. J. M. Stratton, *Mon. Not. R. Astr. Soc.* **71** (1911), 663; also, *Astronomical Physics* (Methuen, 1925), p. 17.

hence
$$\frac{l_1-l_2}{l_2-l_3}=\frac{\lambda_2-\lambda_1}{\lambda_3-\lambda_2}\frac{\lambda_3-Z}{\lambda_1-Z},$$

from which the approximate value Z_0 is found; the value Y_0 is then found from one of the formulae (7) and, thereafter, the value X_0 from any one of the formulae (6).

From (5), with $p=1$, we have

$$a_i\equiv\left(\frac{\partial\phi_i}{\partial X}\right)_0=1,\quad b_i\equiv\left(\frac{\partial\phi_i}{\partial Y}\right)_0=\frac{1}{\lambda_i-Z_0},\quad c_i\equiv\left(\frac{\partial\phi_i}{\partial Z}\right)_0=\frac{Y_0}{(\lambda_i-Z_0)^2}$$

and
$$m_i=l_i-\left(X_0+\frac{Y_0}{\lambda_i-Z_0}\right).$$

The equation of condition is then given by (3).

For a strict determination of the unknowns X_0+x, etc., the set of equations (3) would be treated by the rigorous methods to be described later in this chapter; in some practical problems, however, the approximate solutions X_0, Y_0, Z_0 are taken as the definitive values of the unknowns.

The procedure just described can also be applied advantageously to linear equations of condition by selecting convenient values of X_0, Y_0 and Z_0 close to the values obtained by a preliminary approximate solution of three equations of condition. This procedure, which greatly simplifies the subsequent numerical work, is illustrated in § 6·06.

6·03. Equations of different weights

When, for example, the observational conditions vary throughout a series of measures, a measured quantity m_i will be assigned a particular weight, w_i, according to the principles described in the previous chapter. We write the corresponding equation of condition as

$$A_ix+B_iy+C_iz=M_i, \tag{1}$$

in which M_i now denotes the measured quantity.

If x, y and z denote the *true* values of the unknowns and (M_i) the true value of the measured quantity, then, accurately,

$$A_ix+B_iy+C_iz=(M_i). \tag{2}$$

If $-\epsilon_i$, taken in the usual sense, denotes the error of M_i, then $M_i-(M_i)=-\epsilon_i$. Hence, from (2),

$$A_ix+B_iy+C_iz-M_i=\epsilon_i. \tag{3}$$

This is an accurate equation in which ϵ_i is, of course, unknown; the weight is w_i.

By the principles of § 5·06 the equation (3) is reduced to an equation of unit weight by multiplying throughout by $\sqrt{w_i}$. The equation of condition (1), becomes

$$\sqrt{w_i}\,A_i x+\sqrt{w_i}\,B_i y+\sqrt{w_i}\,C_i z=\sqrt{w_i}\,M_i,$$

which is of the form $$a_i x+b_i y+c_i z=m_i \qquad (4)$$

when a_i is written for $\sqrt{w_i}\,A_i$, ..., and m_i for $\sqrt{w_i}\,M_i$. The last equation is now of unit weight.

The best procedure in dealing with equations of condition of different weights is, *at the outset*, to multiply each throughout by the square root of the corresponding weight, thereby reducing the original system of equations of condition to the system (4) in which each equation is of unit weight.

It is essential to notice that an equation of condition (1) must *not* be multiplied by any numerical factor other than $\sqrt{w_i}$; for example, if we multiply a particular equation (1), of weight w_i, arbitrarily by 2, we are immediately assigning a fictitious weight $4w_i$ to that equation. Similarly, if we multiply a particular equation (4), of unit weight, by 2, we are assigning a fictitious weight 4 to that equation.

In the sequel it will be assumed that the equations of condition, if of unequal weights, have been reduced to equations of equal, or unit, weight.

6·04. Normal equations

When the number of unknowns is three, the normal equations are three equations from which the most probable values of the unknowns x, y and z are calculated. These equations will be derived, first, by the principle of least squares and, secondly, according to the concepts of probability.

(i) Let ξ, η, ζ be undetermined values of the unknowns and V_i the residual defined by

$$V_i=a_i\xi+b_i\eta+c_i\zeta-m_i.$$

According to the principle of least squares, the most plausible values of the unknowns are such that $S\equiv\Sigma V_i^2$ is a minimum, that is, that

$$S\equiv\Sigma(a_i\xi+b_i\eta+c_i\zeta-m_i)^2$$

is a minimum. Accordingly,

$$\frac{\partial S}{\partial\xi}=\frac{\partial S}{\partial\eta}=\frac{\partial S}{\partial\zeta}=0.$$

The first, $\partial S/\partial\xi=0$, gives

$$\Sigma a_i(a_i\xi+b_i\eta+c_i\zeta-m_i)=0,$$

or, in the Gaussian notation,

$$\xi[aa]+\eta[ab]+\zeta[ac]=[am]. \tag{1}$$

Similarly, $\partial S/\partial\eta=0$ and $\partial S/\partial\zeta=0$ give, respectively,

$$\xi[ba]+\eta[bb]+\zeta[bc]=[bm] \tag{2}$$

and

$$\xi[ca]+\eta[cb]+\zeta[cc]=[cm]. \tag{3}$$

It is evident that

$$[ba]\equiv[ab],\quad [ca]\equiv[ac],\quad [cb]\equiv[bc]. \tag{4}$$

The equations (1), (2) and (3) are the *normal equations* from which the most plausible values of the unknowns are obtained by any of the usual methods of solving simultaneous linear equations. If these values are x_0, y_0 and z_0, then

$$\left.\begin{aligned} x_0[aa]+y_0[ab]+z_0[ac]&=[am],\\ x_0[ba]+y_0[bb]+z_0[bc]&=[bm],\\ x_0[ca]+y_0[cb]+z_0[cc]&=[cm].\end{aligned}\right\} \tag{5}$$

(ii) The accurate form of an equation of condition of unit weight is

$$a_i x+b_i y+c_i z-m_i=\epsilon_i, \tag{6}$$

where x, y and z are the true values of the unknowns and ϵ_i is the error associated with m_i.

If h is the modulus of precision of the errors, the probability of making an error between $\epsilon_i-\frac{1}{2}k$ and $\epsilon_i+\frac{1}{2}k$, where k is a small quantity representing the degree of accuracy of the measuring apparatus, is $\dfrac{hk}{\sqrt{\pi}}\exp\{-h^2\epsilon_i^2\}$. The probability, P, that all the errors occur is given by

$$P=C\exp\{-h^2\Sigma\epsilon_i^2\},$$

where

$$C=\left(\frac{hk}{\sqrt{\pi}}\right)^n;$$

hence, by (6),

$$P=C\exp\{-h^2\Sigma(a_i x+b_i y+c_i z-m_i)^2\}.$$

The most probable values of the unknowns are those which make P a maximum or which make $E\equiv\Sigma(a_i x+b_i y+c_i z-m_i)^2$ a minimum. Then $\dfrac{\partial E}{\partial x}=\dfrac{\partial E}{\partial y}=\dfrac{\partial E}{\partial z}=0$; taking first $\dfrac{\partial E}{\partial x}=0$, we have

$$\Sigma a_i(a_i x+b_i y+c_i z-m_i)=0$$

or

$$x[aa]+y[ab]+z[ac]=[am]. \tag{7}$$

Similarly, $\partial E/\partial y = 0$ and $\partial E/\partial z = 0$ give, respectively,

$$x[ba]+y[bb]+z[bc]=[bm] \tag{8}$$

and

$$x[ca]+y[cb]+z[cc]=[cm]. \tag{9}$$

We refer to the group of equations (7), (8) and (9) as the normal equations the solutions of which are x_0, y_0 and z_0. The group just referred to is the same as the group (1), (2) and (3) when ξ, η and ζ are replaced by x, y and z, and as the group (5).

6·05. Checks in forming normal equations

In most problems the formation of the normal equations involves considerable arithmetical calculations, and it is then imperative to apply checks as the work proceeds.

Let s_i be defined by

$$a_i+b_i+c_i+m_i=s_i. \tag{1}$$

Thus, for a particular equation the value of s_i is readily found. From (1), by multiplying by a_i, we have

$$a_ia_i+a_ib_i+a_ic_i+a_im_i=a_is_i,$$

and hence, by summing,

$$[aa]+[ab]+[ac]+[am]=[as]. \tag{2}$$

$[as]$ is formed at the same time as $[aa]$, ..., $[am]$ are formed. Since all the quantities on the left-hand side of (2) all occur in the formation of the first normal equation, (2) affords a simple check on the work up to this point.

From (1) we obtain by a similar process

$$[ba]+[bb]+[bc]+[bm]=[bs], \tag{3}$$

$$[ca]+[cb]+[cc]+[cm]=[cs], \tag{4}$$

and

$$[ma]+[mb]+[mc]+[mm]=[ms], \tag{5}$$

the last identity (5) being a check on the calculation of $[mm]$ which is required later.

6·06. Example of deriving normal equations (Gauss)

To illustrate the process of deriving normal equations, we take Gauss's well-known set of four equations of condition in three unknowns. The equations, of unit weight, are

$$\left.\begin{aligned} X-\ Y+2Z&=\ 3,\\ 3X+2Y-5Z&=\ 5,\\ 4X+\ Y+4Z&=21,\\ -X+3Y+3Z&=14. \end{aligned}\right\} \tag{1}$$

If the first three equations are solved by any elementary method, the results are

$$X = 2\tfrac{4}{7}, \quad Y = 3\tfrac{2}{7}, \quad Z = 1\tfrac{6}{7}.$$

It is easily seen that these values do not quite satisfy the fourth equation.

As mentioned at the end of §6·02 the subsequent arithmetical work is simplified by writing the equations in terms of approximate values of the unknowns; in this case, write

$$X = 2 + x, \quad Y = 3 + y, \quad Z = 2 + z,$$

and then the four equations of condition become

$$\left.\begin{aligned} x - y + 2z &= 0, \\ 3x + 2y - 5z &= 3, \\ 4x + y + 4z &= 2, \\ -x + 3y + 3z &= 1. \end{aligned}\right\} \tag{2}$$

The numbers, m_i, on the right-hand sides of these equations are much simpler than the numbers on the right-hand sides of the original equations (1) in X, Y and Z, an advantage in similar problems when the values of m_i have not the numerical simplicity as in (1).

The four equations of condition, (2), are represented schematically as follows, the corresponding values of s (defined by 6·05 (1)) being given in the final column:

a	b	c	m	s
1	−1	2	0	2
3	2	−5	3	3
4	1	4	2	11
−1	3	3	1	6

(A)

First normal equation. Multiply each of the equations, represented in (A), throughout by the appropriate coefficient of x, that is, by the appropriate value of a in the first column, the same operation being applied to the last column of (A). The results are represented as follows:

aa	ab	ac	am	as
1	−1	2	0	2
9	6	−15	9	9
16	4	16	8	44
1	−3	− 3	−1	−6
27	**6**	**0**	**16**	**49**

(B)

The sums of the several columns are given in the last row; thus, $[aa]=27$, $[ab]=6$, $[ac]=0$ and $[am]=16$; also, $[as]=49$.

The work is checked up to this point by noting that

$$27+6+0+16\equiv 49=[as],$$

in accordance with 6·05 (2).

The first normal equation is

$$27x+6y=16. \tag{3}$$

Second normal equation. Multiply each of the rows in (A) by the appropriate value of b. The results are as follows (we need not fill up the details in the first column since $[ba]\equiv[ab]=6$ from the work in (B), second column).

ba	bb	bc	bm	bs
—	1	−2	0	−2
—	4	−10	6	6
—	1	4	2	11
—	9	9	3	18
6	**15**	**1**	**11**	**33**

(C)

The work is checked since $6+15+1+11\equiv 33=[bs]$.

The second normal equation is

$$6x+15y+z=11. \tag{4}$$

We note: $[bb]=15$, $[bc]=1$, $[bm]=11$.

Third normal equation. Multiply each of the rows in (A) by the appropriate value of c. The results are as follows:

ca	cb	cc	cm	cs
—	—	4	0	4
—	—	25	−15	−15
—	—	16	8	44
—	—	9	3	18
0	**1**	**54**	**−4**	**51**

(D)

We need not fill up the details in the first two columns since $[ca]\equiv[ac]=0$ from (B) and $[cb]\equiv[bc]=1$ from (C); there is the usual check.

The third normal equation is

$$y+54z=-4. \tag{5}$$

We note: $[cc]=54$, $[cm]=-4$.

Calculation of [*mm*]. This quantity occurs in subsequent formulae and it is convenient to evaluate it at this stage. Multiply each of the rows in (A) by the appropriate value of m; we then have

ma	*mb*	*mc*	*mm*	*ms*
—	—	—	0	0
—	—	—	9	9
—	—	—	4	22
—	—	—	1	6
16	**11**	**−4**	**14**	**37**

(E)

We note: $$[mm]=14.$$

Since $[ma]\equiv[am]$, etc., the sums in the first three columns are obtained at once from (B), (C) and (D) respectively; there is the usual check.

Solutions. The three normal equations are:

$$\left.\begin{aligned} 27x+6y\qquad &= 16,\\ 6x+15y+\ z &= 11,\\ y+54z &= -4.\end{aligned}\right\} \tag{F}$$

The solutions, obtained by the usual elementary methods, are

$$x_0=\frac{9{,}356}{19{,}899},\quad y_0=\frac{406}{737}\equiv\frac{406\times27}{19{,}899},\quad z_0=-\frac{1{,}677}{19{,}899}, \tag{6}$$

or, in decimal notation,

$$x_0=0\cdot4702,\quad y_0=0\cdot5509,\quad z_0=-0\cdot0843. \tag{7}$$

The most probable values of the unknowns in the original group of equations (1) are

$$X=2\cdot4702,\quad Y=3\cdot5509,\quad Z=1\cdot9157. \tag{8}$$

6·07. Residuals

It is assumed that the normal equations in § 6·04 have been solved, the values obtained being denoted by x_0, y_0 and z_0; these are the most probable values of the unknowns.

Corresponding to a particular equation of condition, the residual v_i is defined by

$$v_i=a_ix_0+b_iy_0+c_iz_0-m_i. \tag{1}$$

Multiply (1) by a_i and sum; then

$$[av]=x_0[aa]+y_0[ab]+z_0[ac]-[am].$$

But, by 6·04 (5), the right-hand side vanishes; hence $[av]=0$. Similar results are obtained when (1) is multiplied by b_i and the sum taken, and when (1) is multiplied by c_i and the sum taken. We have the group of results:

$$[av]=0, \quad [bv]=0, \quad [cv]=0. \tag{2}$$

Again, multiply (1) by v_i and sum; then

$$[vv]=x_0[av]+y_0[bv]+z_0[cv]-[vm],$$

or, by (2),

$$[vv]=-[vm]. \tag{3}$$

Finally, multiply (1) by m_i and sum; then, by means of (3),

$$-[vv]\equiv[vm]=x_0[am]+y_0[bm]+z_0[cm]-[mm]. \tag{4}$$

Since all the quantities on the right-hand side of (4), x_0, y_0, z_0, $[am]$, ..., $[mm]$, are evaluated in the course of deriving and solving the normal equations, the calculation of $[mm]$ as in 6·06 (E) being included, the sum of the squares of the residuals, that is, $[vv]$, can be calculated in one single operation.

The individual residuals, v_i, can of course be calculated by means of (1) and then $[vv]$ can be obtained. This procedure, in general, entails heavier arithmetical work than in the calculation of $[vv]$ by (4); but it has the advantage of bringing to light an exceptionally large residual, if such exists, thus suggesting a scrutiny of the corresponding measure in case some mistake has been made.

6·08. Gauss's example (evaluation of $[vv]$)

From the calculations in § 6·06 it has been found that

$$[am]=16, \quad [bm]=11, \quad [cm]=-4, \quad [mm]=14;$$

$$x_0=\frac{9{,}356}{19{,}899}, \quad y_0=\frac{406\times 27}{19{,}899}, \quad z_0=-\frac{1{,}677}{19{,}899}.$$

By means of 6·07 (4) it is readily found that

$$[vv]=\frac{1{,}600}{19{,}899}=0{\cdot}0804. \tag{1}$$

Alternatively, on writing $p\equiv 1/19{,}899$, the residual of the first equation in 6·06 (2) is given by

$$v_1\equiv x_0-y_0+2z_0-0=-4960p.$$

Similarly, $v_2=-1320p$, $v_3=1880p$ and $v_4=-1400p$.

From these, $[vv]=1600p=0{\cdot}0804$, as in (1).

If, in any example, $[vv]$ is calculated only by means of the individual residuals, a check is afforded by one of the relations in 6·07 (2), say, $[av]=0$. In the present instance it is easily seen that

$$[av] \equiv 1.v_1 + 3.v_2 + 4.v_3 - 1.v_4 \equiv 0.$$

6·09. Formal solution of the normal equations

We continue to deal with equations of condition in three unknowns. For simplicity, in this case, the following notation is convenient:

$$[aa]=\mathrm{a}, \quad [bb]=\mathrm{b}, \quad [cc]=\mathrm{c};$$

$$[bc]=\mathrm{f}, \quad [ac]=\mathrm{g}, \quad [ab]=\mathrm{h};$$

$$[am]=M_1, \quad [bm]=M_2, \quad [cm]=M_3, \quad [mm]=M.$$

In this notation a, b, ..., h are in roman type to distinguish them from *a*, *b*, ..., *h* in italic type, these latter referring to the constants in the equations of condition.

Since x_0, y_0 and z_0 satisfy the normal equations we have, with the new notation,

$$\mathrm{a}x_0 + \mathrm{h}y_0 + \mathrm{g}z_0 = M_1, \tag{1}$$

$$\mathrm{h}x_0 + \mathrm{b}y_0 + \mathrm{f}z_0 = M_2, \tag{2}$$

$$\mathrm{g}x_0 + \mathrm{f}y_0 + \mathrm{c}z_0 = M_3. \tag{3}$$

Let Δ be defined by

$$\Delta = \begin{vmatrix} \mathrm{a}, & \mathrm{h}, & \mathrm{g} \\ \mathrm{h}, & \mathrm{b}, & \mathrm{f} \\ \mathrm{g}, & \mathrm{f}, & \mathrm{c} \end{vmatrix}. \tag{4}$$

Let A, H and G be the co-factors of a, h and g in the first row of the determinant; then

$$A = \mathrm{bc} - \mathrm{f}^2, \quad H = \mathrm{fg} - \mathrm{ch}, \quad G = \mathrm{fh} - \mathrm{bg}. \tag{5}$$

Similarly,

$$B = \mathrm{ca} - \mathrm{g}^2, \quad C = \mathrm{ab} - \mathrm{h}^2, \quad F = \mathrm{gh} - \mathrm{af}. \tag{6}$$

We have the following well-known properties of the determinant:

$$\mathrm{a}A + \mathrm{h}H + \mathrm{g}G = \Delta, \tag{7}$$

$$\mathrm{h}A + \mathrm{b}H + \mathrm{f}G = 0, \tag{8}$$

$$\mathrm{g}A + \mathrm{f}H + \mathrm{c}G = 0. \tag{9}$$

The left-hand side of (7) is formed from the elements of the first row and their corresponding co-factors; the left-hand sides of (8) and (9) are formed from the elements of the second and third rows and the co-factors of the elements of the first row.

There are two other groups similar to the group (7), (8) and (9) and formed in a similar way.

Multiply (1) by A, (2) by H and (3) by G, and add. The coefficient of x_0 is Δ, by (7); the coefficients of y_0 and z_0 vanish, by (8) and (9); then

$$\Delta x_0 = AM_1 + HM_2 + GM_3. \tag{10}$$

Similarly,

$$\Delta y_0 = HM_1 + BM_2 + FM_3, \tag{11}$$

$$\Delta z_0 = GM_1 + FM_2 + CM_3. \tag{12}$$

The solutions of the normal equations are given by (10), (11) and (12).

From 6·07 (4) the sum of the squares of the residuals is given by

$$[vv] = M - x_0 M_1 - y_0 M_2 - z_0 M_3. \tag{13}$$

In determinant notation the solutions of (1), (2) and (3) are

$$\Delta x_0 = \begin{vmatrix} \mathrm{h}, & \mathrm{g}, & M_1 \\ \mathrm{b}, & \mathrm{f}, & M_2 \\ \mathrm{f}, & \mathrm{c}, & M_3 \end{vmatrix}, \quad \Delta y_0 = - \begin{vmatrix} \mathrm{a}, & \mathrm{g}, & M_1 \\ \mathrm{h}, & \mathrm{f}, & M_2 \\ \mathrm{g}, & \mathrm{c}, & M_3 \end{vmatrix},$$

$$\Delta z_0 = \begin{vmatrix} \mathrm{a}, & \mathrm{h}, & M_1 \\ \mathrm{h}, & \mathrm{b}, & M_2 \\ \mathrm{g}, & \mathrm{f}, & M_3 \end{vmatrix}. \tag{14}$$

Hence (13) can be written

$$\Delta[vv] = \begin{vmatrix} \mathrm{a}, & \mathrm{h}, & \mathrm{g}, & M_1 \\ \mathrm{h}, & \mathrm{b}, & \mathrm{f}, & M_2 \\ \mathrm{g}, & \mathrm{f}, & \mathrm{c}, & M_3 \\ M_1, & M_2, & M_3, & M \end{vmatrix}. \tag{15}$$

Returning to (10) and replacing M_1 by $[am]$, etc., we have

$$\begin{aligned} \Delta x_0 &= A[am] + H[bm] + G[cm] \\ &\equiv A\Sigma a_i m_i + H\Sigma b_i m_i + G\Sigma c_i m_i \\ &= \Sigma m_i(a_i A + b_i H + c_i G). \end{aligned}$$

Define α_i by

$$\Delta\alpha_i = a_i A + b_i H + c_i G; \tag{16}$$

then

$$x_0 = \Sigma m_i \alpha_i = [\alpha m].$$

Similarly, define β_i and γ_i by

$$\Delta\beta_i = a_i H + b_i B + c_i F \tag{17}$$

and

$$\Delta\gamma_i = a_i G + b_i F + c_i C; \tag{18}$$

then we obtain $\quad y_0 = [\beta m] \quad \text{and} \quad z_0 = [\gamma m].$

The formal solutions of the normal equations in terms of the auxiliary quantities α, β and γ are

$$x_0=[\alpha m],\quad y_0=[\beta m],\quad z_0=[\gamma m]. \tag{19}$$

6·10. Properties of α, β and γ

(i) From 6·09 (16),

$$\begin{aligned}\Delta^2\alpha_i^2=&A(Aa_i^2+Ha_ib_i+Ga_ic_i)\\&+H(Aa_ib_i+Hb_i^2+Gb_ic_i)\\&+G(Aa_ic_i+Hb_ic_i+Gc_i^2);\end{aligned}$$

hence, on summing and remembering that $[aa]=\mathrm{a}$, etc., we have

$$\begin{aligned}\Delta^2[\alpha\alpha]=&A(\mathrm{a}A+\mathrm{h}H+\mathrm{g}G)\\&+H(\mathrm{h}A+\mathrm{b}H+\mathrm{f}G)\\&+G(\mathrm{g}A+\mathrm{f}H+\mathrm{c}G)\\=&\Delta A\end{aligned}$$

by means of the relations 6·09 (7), (8) and (9). Hence

$$[\alpha\alpha]=\frac{A}{\Delta}.$$

Similarly, $$[\beta\beta]=\frac{B}{\Delta},\quad [\gamma\gamma]=\frac{C}{\Delta}.$$

(ii) From 6·09 (16) and (17),

$$\begin{aligned}\Delta^2\alpha_i\beta_i=&A(Ha_i^2+Ba_ib_i+Fa_ic_i)\\&+H(Ha_ib_i+Bb_i^2+Fb_ic_i)\\&+G(Ha_ic_i+Bb_ic_i+Fc_i^2),\end{aligned}$$

and hence $$\begin{aligned}\Delta^2[\alpha\beta]=&A(\mathrm{a}H+\mathrm{h}B+\mathrm{g}F)\\&+H(\mathrm{h}H+\mathrm{b}B+\mathrm{f}F)\\&+G(\mathrm{g}H+\mathrm{f}B+\mathrm{c}F).\end{aligned}$$

The coefficient of H in the second term is formed from the elements of the second row of the determinant Δ and the corresponding co-factors, and thus its value is Δ; the coefficients of A and G are zero. Hence

$$[\alpha\beta]=\frac{H}{\Delta}.$$

Similarly, $$[\beta\gamma]=\frac{F}{\Delta},\quad [\alpha\gamma]=\frac{G}{\Delta}.$$

(iii) Multiply 6·09 (16), namely,

$$\Delta\alpha_i = a_i A + b_i H + c_i G, \tag{1}$$

by a_i and sum; then $\Delta[a\alpha] = \mathrm{a}A + \mathrm{h}H + \mathrm{g}G$

$$= \Delta, \quad \text{by } 6{\cdot}09\,(7).$$

Hence $[a\alpha] = 1.$

Similarly, $[b\beta] = 1, \quad [c\gamma] = 1.$

(iv) Multiply (1) by b_i and sum; then

$$\Delta[b\alpha] = \mathrm{h}A + \mathrm{b}H + \mathrm{f}G,$$

and hence $[b\alpha] = 0, \quad$ by 6·09 (8).

Similarly, $[a\beta] = [a\gamma] = [b\gamma] = [c\alpha] = [c\beta] = 0.$

(v) Multiply (1) by v_i and sum; then

$$\Delta[\alpha v] = A[av] + H[bv] + G[cv]$$

$$= 0, \quad \text{by } 6{\cdot}07\,(2).$$

Similarly, $[\beta v] = 0, \quad [\gamma v] = 0.$

6·11. Summary of formulae

The principal formulae derived in the preceding sections are collected here for reference.

$$x_0 = [\alpha m], \quad y_0 = [\beta m], \quad z_0 = [\gamma m]; \tag{1}$$

$$[av] = [bv] = [cv] = 0; \tag{2}$$

$$[\alpha\alpha] = \frac{A}{\Delta}, \quad [\beta\beta] = \frac{B}{\Delta}, \quad [\gamma\gamma] = \frac{C}{\Delta}; \tag{3}$$

$$[\beta\gamma] = \frac{F}{\Delta}, \quad [\alpha\gamma] = \frac{G}{\Delta}, \quad [\alpha\beta] = \frac{H}{\Delta}; \tag{4}$$

$$[a\alpha] = [b\beta] = [c\gamma] = 1; \tag{5}$$

$$[a\beta] = [a\gamma] = [b\alpha] = [b\gamma] = [c\alpha] = [c\beta] = 0; \tag{6}$$

$$[\alpha v] = [\beta v] = [\gamma v] = 0; \tag{7}$$

$$[vv] = [mm] - x_0[am] - y_0[bm] - z_0[cm]. \tag{8}$$

6·12. The formula $\mu^2 = \dfrac{[vv]}{n-3}$

The typical linear equation of condition in three unknowns, expressed accurately in terms of the error ϵ_i associated with the measured quantity, m_i, is, by 6·04 (6),

$$a_i x + b_i y + c_i z - m_i = \epsilon_i, \tag{1}$$

in which x, y and z are the *true* values of the unknowns. The corresponding residual, v_i, is given by

$$a_i x_0 + b_i y_0 + c_i z_0 - m_i = v_i, \tag{2}$$

in which x_0, y_0 and z_0 are the values obtained by solving the normal equations.

(i) Multiply (1) and (2) by α_i and subtract, then

$$a_i\alpha_i(x-x_0) + b_i\alpha_i(y-y_0) + c_i\alpha_i(z-z_0) = \alpha_i\epsilon_i - \alpha_i v_i.$$

By addition of all such equations we obtain

$$(x-x_0)[a\alpha] + (y-y_0)[b\alpha] + (z-z_0)[c\alpha] = [\alpha\epsilon] - [\alpha v],$$

which becomes, by means of 6·11 (5), (6) and (7),

$$x - x_0 = [\alpha\epsilon]. \tag{3}$$

Similarly,

$$y - y_0 = [\beta\epsilon], \quad z - z_0 = [\gamma\epsilon]. \tag{4}$$

(ii) Multiply (1) by v_i and sum; then

$$x[av] + y[bv] + z[cv] - [mv] = [\epsilon v],$$

which becomes, by 6·11 (2),

$$[mv] = -[\epsilon v]. \tag{5}$$

Multiply (2) by v_i and sum; then

$$x_0[av] + y_0[bv] + z_0[cv] - [mv] = [vv],$$

or

$$[vv] = -[mv],$$

or, by (5),

$$[vv] = [\epsilon v]. \tag{6}$$

(iii) Multiply (1) and (2) by ϵ_i and subtract; then sum; the result is

$$(x-x_0)[a\epsilon] + (y-y_0)[b\epsilon] + (z-z_0)[c\epsilon] = [\epsilon\epsilon] - [\epsilon v].$$

By (3), (4) and (6) this last equation becomes

$$[\epsilon\epsilon] = [vv] + [a\epsilon][\alpha\epsilon] + [b\epsilon][\beta\epsilon] + [c\epsilon][\gamma\epsilon]. \tag{7}$$

This is an equation which unites errors with residuals.

Let μ be the standard error of $\epsilon_1, \epsilon_2, \ldots, \epsilon_n$; then

$$\mu^2 = \frac{1}{n}\Sigma\epsilon_i^2$$

or

$$[\epsilon\epsilon] = n\mu^2.$$

We can now write (7) as

$$n\mu^2 = [vv] + P + Q + R, \tag{8}$$

where $P \equiv [a\epsilon][\alpha\epsilon]$, Q and R being defined in a similar way. Now

$$\begin{aligned} P &= (a_1\epsilon_1 + a_2\epsilon_2 + \ldots + a_n\epsilon_n)(\alpha_1\epsilon_1 + \alpha_2\epsilon_2 + \ldots + \alpha_n\epsilon_n) \\ &= \Sigma a_i\alpha_i\epsilon_i^2 + \sum_i\sum_j a_i\alpha_j\epsilon_i\epsilon_j \quad (j \neq i). \end{aligned} \tag{9}$$

This is an exact formula which is susceptible of further simplification only if recourse is had to the general ideas of probability.

Now, the errors ϵ_i, ϵ_j can be positive or negative and, consequently, if n is large the double summation in (9) may be expected to be negligible in comparison with the first summation; otherwise expressed, the probable value of the double summation is zero.

As regards the first summation, $\Sigma a_i\alpha_i\epsilon_i^2$, the best estimate we can make is to take the probable or mean value of each of the quantities ϵ_i^2; this value for an error ϵ is

$$\frac{h}{\sqrt{\pi}}\int_{-\infty}^{\infty}\epsilon^2 e^{-h^2\epsilon^2}d\epsilon \quad \text{or} \quad \frac{1}{2h^2}, \quad \text{by 1·14 (21)},$$

where h is the modulus associated with the errors. Since $\mu^2 = 1/(2h^2)$, the probable value of the first summation in (9) is $\mu^2\Sigma a_i\alpha_i \equiv \mu^2[a\alpha]$. But, by 6·11 (5), $[a\alpha] = 1$; hence the probable value of P is μ^2. The same result follows for the probable values of Q and of R.

Replace P, Q and R in (8) by their probable values; then

$$n\mu^2 = [vv] + 3\mu^2$$

or

$$\mu^2 = \frac{[vv]}{n-3}. \tag{10}$$

As we have seen, the value of $[vv]$ can be found by means of the formula 6·11 (8), or alternatively, by finding the value of each residual by means of 6·07 (1).

From (10) we obtain the standard error, μ, associated with the errors in the measured quantities m.

It is evident that the formula (10) can be generalized when the number of unknowns is k; then

$$\mu^2 = \frac{[vv]}{n-k}. \tag{11}$$

It will be noticed that if $k=1$ we obtain the formula for one unknown derived in § 4·05.

If r denotes the probable error associated with the errors in the measured quantities m then, since $r=0{\cdot}6745\mu$,

$$r=0{\cdot}6745\sqrt{\left\{\frac{[vv]}{n-k}\right\}}, \tag{12}$$

when k is the number of unknowns.

6·13. Precision of x_0, y_0 and z_0

From 6·11 (1),

$$x_0=[\alpha m]=\alpha_1 m_1+\alpha_2 m_2+\ldots+\alpha_n m_n.$$

As in the previous section, h is the modulus of precision for the measures m which, it is to be remembered, are all of unit weight.

Let H be the modulus of precision of x_0; then, by the theorem of § 3·07,

$$\frac{1}{H^2}=\frac{\alpha_1^2}{h^2}+\frac{\alpha_2^2}{h^2}+\ldots+\frac{\alpha_n^2}{h^2}=\frac{[\alpha\alpha]}{h^2}. \tag{1}$$

Let μ_x be the standard error of x_0 so that $\mu_x^2=1/(2H^2)$; then, since $\mu^2=1/(2h^2)$, (1) becomes

$$\mu_x^2=\mu^2[\alpha\alpha]$$

$$=\mu^2\frac{A}{\Delta}, \quad \text{by 6·11 (3).}$$

By analogy with 5·02 (8) we write

$$\mu_x^2=\frac{\mu^2}{w_x};$$

then $$w_x=\frac{\Delta}{A}\equiv\frac{1}{[\alpha\alpha]},$$

and w_x is the *weight* of x_0.

Hence the precision of x_0, which is represented by the standard error μ_x, is given by

$$\mu_x=\frac{\mu}{\sqrt{w_x}}=\sqrt{\left\{\frac{[vv]}{(n-3)\,w_x}\right\}}. \tag{2}$$

The weights, w_y and w_z, of y_0 and z_0 are defined in a similar way; they are

$$w_y=\frac{\Delta}{B}\equiv\frac{1}{[\beta\beta]}$$

and $$w_z=\frac{\Delta}{C}\equiv\frac{1}{[\gamma\gamma]}.$$

The standard errors, μ_y and μ_z, of y_0 and z_0 are given by formulae similar to (2).

The collected results are:

$$\mu_x = \frac{\mu}{\sqrt{w_x}}, \quad \mu_y = \frac{\mu}{\sqrt{w_y}}, \quad \mu_z = \frac{\mu}{\sqrt{w_z}}; \tag{3}$$

$$w_x = \frac{\Delta}{A}, \quad w_y = \frac{\Delta}{B}, \quad w_z = \frac{\Delta}{C}; \tag{4}$$

$$\mu = \sqrt{\left\{\frac{[vv]}{n-3}\right\}}. \tag{5}$$

If the precision of x_0, for example, is expressed in terms of the probable error, r_x, then

$$r_x \equiv 0{\cdot}6745\mu_x = 0{\cdot}6745\frac{\mu}{\sqrt{w_x}}$$

or

$$r_x = 0{\cdot}6745\sqrt{\left\{\frac{[vv]}{(n-3)\,w_x}\right\}}, \tag{6}$$

with similar formulae for r_y and r_z.

6·14. Gauss's example (precision of solutions)

From 6·06 (F), the normal equations are

$$\begin{aligned} 27x + 6y \qquad &= 16, \\ 6x + 15y + z &= 11, \\ y + 54z &= -4. \end{aligned}$$

The determinant, Δ, is given by

$$\Delta = \begin{vmatrix} 27, & 6, & 0 \\ 6, & 15, & 1 \\ 0, & 1, & 54 \end{vmatrix} = 27\begin{vmatrix} 15, & 1 \\ 1, & 54 \end{vmatrix} - 6\begin{vmatrix} 6, & 1 \\ 0, & 54 \end{vmatrix}$$

or

$$\Delta = 19{,}899.$$

Also, $$A = \begin{vmatrix} 15, & 1 \\ 1, & 54 \end{vmatrix}, \quad B = \begin{vmatrix} 27, & 0 \\ 0, & 54 \end{vmatrix}, \quad C = \begin{vmatrix} 27, & 6 \\ 6, & 15 \end{vmatrix}$$

or

$$A = 809, \quad B = 1458, \quad C = 369.$$

The weights to be assigned to x_0, y_0 and z_0 are given by

$$w_x \equiv \frac{\Delta}{A} = 24{\cdot}6, \quad w_y \equiv \frac{\Delta}{B} = 13{\cdot}6, \quad w_z \equiv \frac{\Delta}{C} = 53{\cdot}9. \tag{1}$$

From 6·08 (1), $[vv] = 0{\cdot}0804$; hence, since the number, n, of equations of condition is 4,

$$\mu^2 \equiv \frac{[vv]}{n-3} = 0{\cdot}0804,$$

from which

$$\mu = 0{\cdot}2835. \tag{2}$$

Then,

$$\mu_x \equiv \frac{\mu}{\sqrt{w_x}} = 0{\cdot}057, \quad \mu_y = 0{\cdot}077, \quad \mu_z = 0{\cdot}039. \tag{3}$$

The probable errors r_x, etc., are, numerically,

$$r_x = 0{\cdot}039, \quad r_y = 0{\cdot}052, \quad r_z = 0{\cdot}026. \tag{4}$$

The most probable values of the unknowns X, Y and Z in the equations of condition 6·06 (1) are given by 6·06 (8). The final solutions are written (to 3 places of decimals)

$$\left.\begin{aligned} X &= 2{\cdot}470 \pm 0{\cdot}057 \\ Y &= 3{\cdot}551 \pm 0{\cdot}077 \\ Z &= 1{\cdot}916 \pm 0{\cdot}039 \end{aligned}\right\} \quad (\text{S.E.}), \tag{5}$$

in which the precision is expressed in terms of standard error, or

$$\left.\begin{aligned} X &= 2{\cdot}470 \pm 0{\cdot}039 \\ Y &= 3{\cdot}551 \pm 0{\cdot}052 \\ Z &= 1{\cdot}916 \pm 0{\cdot}026 \end{aligned}\right\} \quad (\text{P.E.}), \tag{6}$$

in which the precision is expressed in terms of probable error.

In (5) the standard error of Z is the least of the three, from which it is inferred that the determination of Z is to be considered more reliable than that of X and of Y.

If the four equations of condition in 6·06 (1) referred to the results of *real* observations, the smallness of the standard errors (or the probable errors) given above suggest that the measures have been made with a high degree of accuracy, so that considerable confidence can be placed in the most probable values of the unknowns X, Y and Z derived from the normal equations.

The preceding calculations, depending on the evaluation of determinants, are easy and straightforward, due to the simple character of the coefficients in the normal equations; Gauss's example has, in fact, been chosen to illustrate general principles and procedure so that the arithmetical processes can be readily followed. The determinant method is not, however, always the most suitable practical method when the coefficients of x, y and z in the normal equations are such numbers as 27·53, 96·82, etc., or, generally, when the number of unknowns is greater than two. The normal equations have to be solved in any event and Gauss's method of solution with which we deal in the next section has the advantage of giving, without additional calculation, the weights to be assigned to two of the unknowns while the calculation of the third weight is extremely simple.

6·15. Gauss's method of solving the normal equations and evaluating the weights of the unknowns

The normal equations for three unknowns are, in the simplified notation introduced in § 6·09,

$$\mathrm{a}x+\mathrm{h}y+\mathrm{g}z=M_1, \tag{1}$$

$$\mathrm{h}x+\mathrm{b}y+\mathrm{f}z=M_2, \tag{2}$$

$$\mathrm{g}x+\mathrm{f}y+\mathrm{c}z=M_3, \tag{3}$$

where $\mathrm{a}\equiv[aa]$, etc., $M_1\equiv[am]$, etc.

The first stages of the solutions are the successive eliminations of x and y.

(i) *Elimination of* x. From (1)

$$x=-\frac{\mathrm{h}}{\mathrm{a}}y-\frac{\mathrm{g}}{\mathrm{a}}z+\frac{M_1}{\mathrm{a}}. \tag{4}$$

Substitute this expression for x in (2); then

$$y\left\{\mathrm{b}-\frac{\mathrm{h}^2}{\mathrm{a}}\right\}+z\left\{\mathrm{f}-\frac{\mathrm{gh}}{\mathrm{a}}\right\}=M_2-\frac{\mathrm{h}}{\mathrm{a}}M_1. \tag{5}$$

Substitute for x in (3); then

$$y\left\{\mathrm{f}-\frac{\mathrm{gh}}{\mathrm{a}}\right\}+z\left\{\mathrm{c}-\frac{\mathrm{g}^2}{\mathrm{a}}\right\}=M_3-\frac{\mathrm{g}}{\mathrm{a}}M_1. \tag{6}$$

The coefficients of y and z in (5) *and* (6) *are to be evaluated as they are written*; for example, these equations are *not* to be 'simplified' by multiplying throughout by a or by any other number.

In Gauss's notation the equations (5) and (6) are written as

$$y[bb1]+z[bc1]=[bm1], \tag{7}$$

$$y[bc1]+z[cc1]=[cm1], \tag{8}$$

where, in terms of the simplified notation and Gauss's bracket notation,

$$[bb1] \equiv \mathrm{b} - \frac{\mathrm{h}^2}{\mathrm{a}} \equiv \frac{C}{\mathrm{a}} = [bb] - \frac{[ab][ab]}{[aa]}, \tag{9}$$

$$[bc1] \equiv \mathrm{f} - \frac{\mathrm{gh}}{\mathrm{a}} \equiv -\frac{F}{\mathrm{a}} = [bc] - \frac{[ab][ac]}{[aa]}, \tag{10}$$

$$[cc1] \equiv \mathrm{c} - \frac{\mathrm{g}^2}{\mathrm{a}} \equiv \frac{B}{\mathrm{a}} = [cc] - \frac{[ac][ac]}{[aa]}, \tag{11}$$

$$[bm1] \equiv M_2 - \frac{\mathrm{h}}{\mathrm{a}} M_1 = [bm] - \frac{[ab][am]}{[aa]}, \tag{12}$$

$$[cm1] \equiv M_3 - \frac{\mathrm{g}}{\mathrm{a}} M_1 = [cm] - \frac{[ac][am]}{[aa]}, \tag{13}$$

which are all summarized in the single formula

$$[pq1] = [pq] - \frac{[ap][aq]}{[aa]}, \tag{14}$$

where p and q are any two of b, c and m, or $p=q=b$ or $p=q=c$, the 1 in the new Gaussian (triple) bracket being associated with the first letter, namely, a.

All of the triple brackets are evaluated in the course of the arithmetical work.

In (9), (10) and (11), it is to be remembered that C, F and B are the usual co-factors of the determinant

$$\Delta = \begin{vmatrix} \mathrm{a}, & \mathrm{h}, & \mathrm{g} \\ \mathrm{h}, & \mathrm{b}, & \mathrm{f} \\ \mathrm{g}, & \mathrm{f}, & \mathrm{c} \end{vmatrix}.$$

The formulae for the triple brackets in terms of C, F, etc., will be used when the weights of x, y and z are considered.

(ii) *Elimination of y and calculation of z.* We now deal with equations (7) and (8). From (7)

$$y = -z\frac{[bc1]}{[bb1]} + \frac{[bm1]}{[bb1]}. \tag{15}$$

Substitute this expression for y in (8); then

$$z\left\{[cc1] - \frac{[bc1]^2}{[bb1]}\right\} = [cm1] - \frac{[bc1][bm1]}{[bb1]}, \tag{16}$$

or, in the Gaussian notation,

$$z[cc2] = [cm2], \tag{17}$$

where $[cc2]$ is the *unsimplified* coefficient of z given by

$$[cc2]=[cc1]-\frac{[bc1]^2}{[bb1]}, \tag{18}$$

and $[cm2]$ is written for the right-hand side of (16).

From (17), the value of z is given by

$$z=\frac{[cm2]}{[cc2]}; \tag{19}$$

the numerator and denominator are evaluated in the course of the arithmetical work.

According to our previous notation the solution in (19) is designated z_0; however, here and in the sequel it is sufficient to refer to the three values, obtained in the solution of the normal equations, simply by x, y and z.

(iii) *Weight of z.* From (18), by means of (9), (10) and (11),

$$[cc2]=\frac{B}{\mathrm{a}}-\frac{F^2}{\mathrm{a}C}=\frac{BC-F^2}{\mathrm{a}C}.$$

But
$$BC-F^2=(\mathrm{ac}-\mathrm{g}^2)(\mathrm{ab}-\mathrm{h}^2)-(\mathrm{gh}-\mathrm{af})^2 \tag{20}$$
$$=\mathrm{a}\Delta;$$

hence
$$[cc2]=\frac{\Delta}{C}, \tag{21}$$

or, from 6·13 (4),
$$[cc2]=w_z. \tag{22}$$

Thus the weight, w_z, of z is derived, without additional calculation, as a by-product of the process of elimination and solution.

(iv) *Calculation of y and the weight of y.* From (8),

$$z=-y\frac{[bc1]}{[cc1]}+\frac{[cm1]}{[cc1]}; \tag{23}$$

substitute this expression for z in (7); then

$$y\left\{[bb1]-\frac{[bc1]^2}{[cc1]}\right\}=[bm1]-\frac{[bc1][cm1]}{[cc1]} \tag{24}$$

or
$$y[bb2]=[bm2], \tag{25}$$

in which
$$[bb2]=[bb1]-\frac{[bc1]^2}{[cc1]}, \tag{26}$$

$[bb2]$ being the *unsimplified* coefficient of y in (24).

From (25) the value of y is obtained, the expression for $[bm2]$ being given by the right-hand side of (24).

Now, from (26) and (9), (10) and (11),

$$[bb2] = \frac{C}{\text{a}} - \frac{F^2}{\text{a}B} = \frac{\Delta}{B}, \quad \text{by (20).}$$

Thus

$$[bb2] = w_y, \tag{27}$$

and its value is obtained, without additional calculation, in the process of solution.

(v) *Calculation of x and w_x.* The value of x is obtained by direct substitution of y and z by means of (25) and (19) in (4).

Since $w_x = \Delta/A$, the weight, w_x, can be calculated directly by evaluating the determinant and the co-factor A.

The determinant can be readily evaluated in terms of quantities found in the course of the preceding calculations; thus

$$\Delta \equiv \frac{\Delta}{C}\frac{C}{\text{a}}\,\text{a} = w_z[bb1]\,\text{a}$$

by means of (21), (22) and (9); hence

$$w_x = \frac{w_z[bb1]\,\text{a}}{(\text{bc} - \text{f}^2)}. \tag{28}$$

(vi) *Derivation of the final results.* The standard error, μ, for the equations of condition of equal weight is given by

$$\mu = \sqrt{\left\{\frac{[vv]}{n-3}\right\}}, \tag{29}$$

and $[vv]$ is found, either by means of the formula 6·11 (8) in which x_0, y_0 and z_0 are the values, as found above, representing the solutions of the normal equations or, alternatively, by calculating the individual residuals. The results, in terms of standard errors, for the unknowns are then expressed as follows:

$$x_0 \pm \frac{\mu}{\sqrt{w_x}}, \quad y_0 \pm \frac{\mu}{\sqrt{w_y}}, \quad z_0 \pm \frac{\mu}{\sqrt{w_z}} \quad \text{(S.E.)}, \tag{30}$$

or, in terms of probable errors,

$$x_0 \pm \frac{r}{\sqrt{w_x}}, \quad y_0 \pm \frac{r}{\sqrt{w_y}}, \quad z_0 \pm \frac{r}{\sqrt{w_z}} \quad \text{(P.E.)}, \tag{31}$$

where

$$r = 0{\cdot}6745\sqrt{\left\{\frac{[vv]}{n-3}\right\}}.$$

6·16. Gauss's example (solution by Gauss's method)

In this section Gauss's method, described in the previous section, will be illustrated, using the normal equations derived in § 6·06.

The equations of this section will be numbered to correspond with the equations in § 6·15.

The normal equations are:

$$27x + 6y \qquad = 16, \tag{1}$$

$$6x + 15y + \quad z = 11, \tag{2}$$

$$y + 54z = -4. \tag{3}$$

(i) *Elimination of x.* From (1),

$$x = -\tfrac{2}{9}y + \tfrac{16}{27}. \tag{4}$$

Substitute this expression for x in (2); then

$$y\{15 - \tfrac{4}{3}\} + z = 11 - \tfrac{32}{9}$$

or

$$\tfrac{41}{3}y + z = \tfrac{67}{9}. \tag{5) or (7}$$

Since (3) is independent of x, substitution by means of (4) is unnecessary. We rewrite (3):

$$y + 54z = -4. \tag{6) or (8}$$

Comparing these last two equations with 6·15 (7) and (8), we have the identifications:

$$[bb1] = \tfrac{41}{3}, \quad [bc1] = 1, \quad [cc1] = 54, \quad [bm1] = \tfrac{67}{9}, \quad [cm1] = -4. \tag{A}$$

(ii) *Elimination of y and calculation of z.* From (5) or (7),

$$y = -\tfrac{3}{41}z + \tfrac{67}{123}. \tag{15}$$

Substitute this expression for y in (6) or (8); then

$$z\{54 - \tfrac{3}{41}\} = -4 - \tfrac{67}{123} \tag{16}$$

or

$$\tfrac{2211}{41}z = -\tfrac{559}{123}. \tag{17}$$

Hence

$$z \equiv z_0 = -\tfrac{559}{6633} = -0{\cdot}084. \tag{19}$$

From (17) we have the identifications:

$$[cc2] = \tfrac{2211}{41}, \quad [cm2] = -\tfrac{559}{123}. \tag{18}$$

(iii) *Weight of z.* By 6·15 (22) the weight of z is the coefficient of z in 6·15 (17), namely, $[cc2]$; hence

$$w_z = \tfrac{2211}{41} = 53{\cdot}9. \tag{22}$$

(iv) *Calculation of y and the weight of y.* From (8),

$$z=-\tfrac{1}{54}y-\tfrac{2}{27}. \tag{23}$$

Substitute this expression for z in (7); then

$$y(\tfrac{41}{3}-\tfrac{1}{54})=\tfrac{67}{9}+\tfrac{2}{27}$$

or

$$\tfrac{737}{54}y=\tfrac{203}{27}. \tag{25}$$

We have the identifications:

$$[bb2]=\tfrac{737}{54},\quad [bm2]=\tfrac{203}{27}.$$

Hence, from (25), $\quad y\equiv y_0=\tfrac{406}{737}=0{\cdot}551.$

By 6·15 (27), the weight of y is $[bb2]$ or $\tfrac{737}{54}$; hence

$$w_y=13{\cdot}6. \tag{27}$$

(v) *Calculation of x and the weight of x.* Substitute in (4) the values $\tfrac{406}{737}$ and $-\tfrac{559}{6633}$ for y and z respectively; then

$$x\equiv x_0=\tfrac{9356}{19899}=0{\cdot}470.$$

In 6·15 (28) we have the following values:

$$\mathrm{a}=27,\quad w_z=\tfrac{2211}{41},\quad [bb1]=\tfrac{41}{3},\quad \text{from (A);}$$

and

$$\mathrm{bc}-\mathrm{f}^2=\begin{vmatrix}15, & 1\\ 1, & 54\end{vmatrix}=809.$$

Hence

$$w_x=\tfrac{19899}{809}=24{\cdot}6. \tag{28}$$

(vi) *Results.* From 6·14 (2),

$$\mu=0{\cdot}2835. \tag{29}$$

From (28), (27) and (22),

$$w_x=24{\cdot}6,\quad w_y=13{\cdot}6,\quad w_z=53{\cdot}9.$$

From 6·15 (30),

$$\mu_x\equiv\frac{\mu}{\sqrt{w_x}}=0{\cdot}057,\quad \mu_y=0{\cdot}077,\quad \mu_z=0{\cdot}039\quad (\text{S.E.}),$$

and from 6·15 (31),

$$r_x=0{\cdot}6745\mu_x=0{\cdot}039,\quad r_y=0{\cdot}052,\quad r_z=0{\cdot}026\quad (\text{P.E.}).$$

All these results have been obtained previously in § 6·14.

The final results, relating to the original equations of condition in terms of X, Y and Z, are as given by 6·14 (5) and (6).

6·17. Checks for Gauss's method

(i) A typical equation of condition is

$$a_i x + b_i y + c_i z = m_i.$$

In § 6·05 we showed how checks on the formation of the normal equations are undertaken by introducing a subsidiary quantity s_i defined by

$$a_i + b_i + c_i + m_i = s_i.$$

In the equations of condition a measured quantity m will be given in terms of some unit unrelated to the magnitudes of the coefficients a, b and c. To avoid complications in the computations it is advisable to choose a new unit for m so that the resulting numerical values of m are of the same order of magnitude as for a, b and c, all being expressed to the same number of decimal places.

By 6·05 (2), (3) and (4),

$$[aa]+[ab]+[ac]+[am]=[as], \tag{1}$$

$$[ba]+[bb]+[bc]+[bm]=[bs], \tag{2}$$

$$[ca]+[cb]+[cc]+[cm]=[cs]. \tag{3}$$

It is supposed, as in § 6·06, that $[as]$, $[bs]$ and $[cs]$ are evaluated in providing checks on the formation of the normal equations. We write the normal equations with the addition of $[as]$, $[bs]$, $[cs]$ as follows:

$$x[aa]+y[ab]+z[ac]=[am]; \quad [as], \tag{4}$$

$$x[ba]+y[bb]+z[bc]=[bm]; \quad [bs], \tag{5}$$

$$x[ca]+y[cb]+z[cc]=[cm]; \quad [cs], \tag{6}$$

which are solved, in Gauss's method as in § 6·15 by the successive elimination of x and y, operating on $[as]$ in the same way as on $[am]$, on $[bs]$ as on $[bm]$ and on $[cs]$ as on $[cm]$; the process is in fact the same as solving two sets of equations, first, with $[am]$, etc., on the right-hand sides and, secondly, with $[as]$, etc., on the right-hand sides, the two solutions being carried on side by side. For example, the elimination of x between (1) and (2) results in 6·15 (5) or (7), namely,

$$y[bb1]+z[bc1]=[bm1]; \quad [bs1],$$

in which all the triple brackets are evaluated.

Now, by 6·15 (9), (10) and (12),

$$[bb1]+[bc1]+[bm1]=[bb]+[bc]+[bm]-\frac{[ba]}{[aa]}\{[ab]+[ac]+[am]\}$$

$$=[bs]-[ab]-\frac{[ba]}{[aa]}\{[as]-[aa]\},$$

by means of (2) and (1). The right-hand side is thus

$$[bs]-\frac{[ba][as]}{[aa]} \quad \text{or} \quad [bs1].$$

Hence

$$[bb1]+[bc1]+[bm1]=[bs1], \tag{7}$$

which provides the check up to this point. Similarly,

$$[bc1]+[cc1]+[cm1]=[cs1], \tag{8}$$

which provides the check on 6·15 (8).

Again, from 6·15 (18),

$$[cc2]+[cm2]=[cc1]+[cm1]-\frac{[bc1]}{[bb1]}\{[bc1]+[bm1]\}$$

$$=[cs1]-[bc1]-\frac{[bc1]}{[bb1]}\{[bs1]-[bb1]\}, \quad \text{from (8) and (7)}$$

$$=[cs1]-\frac{[bc1][bs1]}{[bb1]}$$

$$=[cs2],$$

by the analogue of 6·15 (18).

Thus a check is provided for the numerical quantities in 6·15 (17).

(ii) If the individual residuals are evaluated, then $[vv]$ is calculated and this should agree with the value of $[vv]$ obtained by means of the formula 6·11 (8), namely,

$$[vv]=[mm]-x_0[am]-y_0[bm]-z_0[cm].$$

Alternatively, the relation $[av]=0$ in 6·11 (2) may be used for checking the residuals.

(iii) *Application to Gauss's example.* The pseudo-normal equations, in terms of s, are obtained as follows:

from 6·16 (1), $\quad s_1=27+\ 6+16 \quad =49;$

from 6·16 (2), $\quad s_2=\ 6+15+\ 1+11=33;$

from 6·16 (3), $\quad s_3=\ 1+54-\ 4 \quad =51.$

The pseudo-normal equations are

$$27x+\ 6y \qquad =49, \tag{9}$$

$$6x+15y+\ \ z=33, \tag{10}$$

$$y+54z=51.$$

From (9), $\quad x=-\frac{2}{9}y+\frac{49}{27},$

and substitution in (10) gives

$$y(15-\tfrac{4}{3})+z=33-\tfrac{49}{27}.6$$

or

$$\tfrac{41}{3}y+z=\tfrac{199}{9}. \qquad (11)$$

Compare (11) with 6·16 (5) in which $[bm1]=\frac{67}{9}$; thus, from (11),

$$[bs1]=\tfrac{199}{9}. \qquad (12)$$

Now, from (A) in § 6·16 (i),

$$[bb1]+[bc1]+[bm1]=\tfrac{41}{3}+1+\tfrac{67}{9}=\tfrac{199}{9}$$

$$=[bs1], \quad \text{by (12)}.$$

Thus, the check given by (7) shows that the work is correct up to this point.

The remaining checks are applied in a similar way.

6·18. Alternative method of calculating weights

We consider the normal equations in § 6·15; by 6·09 (10) the solution for x is given by

$$x_0=\frac{A}{\Delta}M_1+\frac{H}{\Delta}M_2+\frac{G}{\Delta}M_3.$$

Now $w_x=\Delta/A$; hence if we solve the normal equations, in which we put

$$M_1=1, \quad M_2=0, \quad M_3=0$$

we obtain a value x' which is equal to $1/w_x$. The method can clearly be applied to the evaluation of w_y and w_z, each requiring a separate solution of three quasi-normal equations. Further, the normal equations and the quasi-normal equations can be solved in any way we please.

To illustrate the calculation of w_x in Gauss's example, we have the three quasi-normal equations

$$27x+\ 6y \qquad =1, \qquad (1)$$

$$6x+15y+\ z=0, \qquad (2)$$

$$y+54z=0. \qquad (3)$$

Eliminate z between (2) and (3); then

$$324x+809y=0. \qquad (4)$$

From (1) and (4) it is easily found that $x=809/19899$, so that

$$w_x=\tfrac{19899}{809}=24{\cdot}6,$$

as found in 6·14 (1).

6·19. Equations of condition in two unknowns

As this case is of frequent occurrence, the formal solutions will be given, followed in the next section by a numerical example relating to the photographic determination of the parallax of a star.

A typical equation of condition is

$$a_i x + b_i y = m_i \quad (i=1, 2, \ldots, n).$$

The normal equations, formed in the usual way, are

$$x[aa] + y[ab] = [am],$$

$$x[ba] + y[bb] = [bm],$$

which are written for convenience as

$$\mathrm{a}x + \mathrm{h}y = M_1, \tag{1}$$

$$\mathrm{h}x + \mathrm{b}y = M_2, \tag{2}$$

where $\mathrm{a}=[aa]$, $\mathrm{h}=[ab]$, $\mathrm{b}=[bb]$, $M_1=[am]$, $M_2=[bm]$.

Let Δ denote the determinant given by

$$\Delta = \begin{vmatrix} \mathrm{a}, & \mathrm{h} \\ \mathrm{h}, & \mathrm{b} \end{vmatrix};$$

then, the solutions of (1) and (2) are x_0, y_0 given by

$$\Delta x_0 = \mathrm{b}M_1 - \mathrm{h}M_2, \tag{3}$$

$$\Delta y_0 = \mathrm{a}M_2 - \mathrm{h}M_1. \tag{4}$$

If w_x and w_y are the respective weights of the solutions x_0 and y_0 then, as in § 6·18, $1/w_x$ is the value x' obtained by solving (1) and (2) when M_1 is replaced by 1 and M_2 is replaced by 0. From (3), $x' = \mathrm{b}/\Delta$; hence

$$w_x = \Delta/\mathrm{b}.$$

Similarly, $$w_y = \Delta/\mathrm{a}.$$

The formula for $[vv]$ is

$$[vv] = [mm] - x_0[am] - y_0[bm].$$

Also, the standard error, μ, associated with the equations of condition is given, by 6·12 (11), by

$$\mu^2 = \frac{[vv]}{n-2}.$$

If μ_x, μ_y denote the standard errors of the unknowns, then

$$\mu_x = \sqrt{\left\{\frac{[vv]}{(n-2)\,w_x}\right\}},$$

with a similar equation for μ_y.

The final results are written, first, in terms of standard errors, as

$$x = x_0 \pm \mu_x, \quad y = y_0 \pm \mu_y \quad \text{(S.E.)},$$

and, secondly, in terms of probable errors, r_x and r_y, as

$$x = x_0 \pm r_x, \quad y = y_0 \pm r_y \quad \text{(P.E.)},$$

where
$$r_x \equiv 0{\cdot}6745\mu_x = 0{\cdot}6745 \sqrt{\left\{\frac{[vv]}{(n-2)\,w_x}\right\}}$$

with a similar formula for r_y.

6·20. Example (two unknowns)

As stated in §6·01 the practical problem of determining the parallax, Π, of a star from measures made on pairs of photographic plates involves equations of condition of the form

$$ax + by = m,$$

in which x and y are written for Π and μ_α respectively, m is a measured quantity, and a and b are known numerical constants for a particular pair of plates.

The details, which follow, refer to the measures† on six pairs of plates for the star 'Boss 3650'; there are thus six equations of condition. The quantities a, b and m are given in Table 13, the latter in terms of a unit to be referred to later; x and y will first be found in terms of this unit. It will be observed that a, b and m are all of a similar magnitude, each being given to two places of decimals. Table 13 also includes the values of the quantity $s \equiv a + b + m$, introduced for the purpose of checking the arithmetical work in deriving the normal equations; in the last two columns are the values of the residuals, v, and of v^2 which will be used in the subsequent calculations. Tables 14 and 15 contain the calculations of $[aa]$, etc., and Table 16 contains the calculation of $[mm]$ which, carried to an additional place of decimals, will be required later.

It is seen that the work in Tables 14–16 is checked; in Table 14, for example,

$$[aa] + [ab] + [am] = [as].$$

From Tables 14 and 15, the normal equations are

$$13{\cdot}444x + 1{\cdot}676y = 4{\cdot}731, \tag{1}$$

$$1{\cdot}676x + 1{\cdot}782y = -2{\cdot}275. \tag{2}$$

† A. van Maanen, *Contributions of the Mt. Wilson Solar Observatory*, no. 111 (1915), 17.

Table 13

a	b	m	s	v	v^2
1·50	−0·32	+1·53	2·71	−0·079	0·00624
1·50	−0·32	+1·34	2·52	+0·111	0·01232
1·50	−0·32	+1·50	2·68	−0·049	0·00240
1·42	+0·73	−0·48	1·67	−0·027	0·00073
1·64	+0·64	−0·31	1·97	+0·094	0·00884
1·41	+0·73	−0·45	1·69	−0·063	0·00397
8·97	**+1·14**	**+3·13**	**13·24**	—	**0·03450**

Table 14

aa	ab	am	as
2·250	−0·480	+2·295	4·065
2·250	−0·480	+2·010	3·780
2·250	−0·480	+2·250	4·020
2·016	+1·037	−0·682	2·371
2·690	+1·050	−0·508	3·231
1·988	+1·029	−0·634	2·383
13·444	**+1·676**	**+4·731**	**19·850**

Table 15

ba	bb	bm	bs
—	0·102	−0·490	−0·867
—	0·102	−0·429	−0·806
—	0·102	−0·480	−0·858
—	0·533	−0·350	+1·219
—	0·410	−0·198	+1·261
—	0·533	−0·328	+1·234
1·676	**1·782**	**−2·275**	**+1·183**

Table 16

ma	mb	mm	ms
—	—	2·3409	+4·146
—	—	1·7956	+3·377
—	—	2·2500	+4·020
—	—	0·2304	−0·802
—	—	0·0961	−0·611
—	—	0·2025	−0·760
4·731	**−2·275**	**6·9155**	**+9·370**

The solutions of equations (1), (2) for x_0 and y_0, can be effected in a variety of ways. For simplicity in formal presentation we use the determinant method. Here

$$\Delta = \begin{vmatrix} 13\cdot444, & 1\cdot676 \\ 1\cdot676, & 1\cdot782 \end{vmatrix} = 21\cdot15;$$

then $$\Delta x_0 = 4\cdot731 \times 1\cdot782 + 2\cdot275 \times 1\cdot676,$$

and $$\Delta y_0 = -13\cdot444 \times 2\cdot275 - 1\cdot676 \times 4\cdot731,$$

from which $$x_0 = 0\cdot579, \quad y_0 = -1\cdot821. \tag{3}$$

The residuals, v, can then be calculated by means of the formula

$$v = ax_0 + by_0 - m;$$

the values of v and of v^2 are given in Table 13, from which

$$[vv] = 0\cdot0345. \tag{4}$$

$[vv]$ can also be calculated by means of the formula 6·07 (4), namely,

$$[vv] = [mm] - x_0[am] - y_0[bm]$$
$$= 6\cdot9155 - 4\cdot731 x_0 + 2\cdot275 y_0$$

from the data in Tables 16, 14 and 15. Thus

$$[vv] = 0\cdot0340. \tag{5}$$

The difference between the values in (4) and (5) is due to the curtailment of the numbers of decimal places in the various calculations.

We use (4) to calculate the standard error, μ, of the measured quantities, m. Applying the formula $\mu = \sqrt{\left\{\frac{[vv]}{n-k}\right\}}$, in which $n=6$ (the number of equations of condition) and $k=2$ (the number of unknowns), we obtain

$$\mu = \sqrt{\left\{\frac{0\cdot0345}{4}\right\}} = 0\cdot0929. \tag{6}$$

From 6·13 (4), the weights w_x and w_y of the unknowns are Δ/A and Δ/B, where A and B are the co-factors of $[aa]$ and $[bb]$ respectively in the determinant; from (1) and (2),

$$[aa] = 13\cdot444, \quad [bb] = 1\cdot782;$$

hence $A = 1\cdot782$ and $B = 13\cdot444$ so that

$$w_x = \frac{21\cdot15}{1\cdot782} \quad \text{and} \quad w_y = \frac{21\cdot15}{13\cdot44}.$$

The standard errors, μ_x and μ_y, of the unknowns are

$$\frac{\mu}{\sqrt{w_x}} \quad \text{and} \quad \frac{\mu}{\sqrt{w_y}};$$

hence $\mu_x = 0{\cdot}0270, \quad \mu_y = 0{\cdot}0740.$

The probable errors, r_x and r_y, of the unknowns are then

$$r_x = 0{\cdot}0182, \quad r_y = 0{\cdot}0499.$$

We write the solutions in terms of probable error and in terms of the unit of measurement as

$$\left.\begin{array}{l} x_0 = 0{\cdot}579 \pm 0{\cdot}0182 \\ y_0 = -1{\cdot}821 \pm 0{\cdot}0499 \end{array}\right\} \text{(P.E.)}.$$

The unit of measurement is equivalent to $0''{\cdot}1649$. The final results in terms of probable error, are, to three decimal places,

$$\left.\begin{array}{l} \Pi \equiv x_0 = 0''{\cdot}096 \pm 0''{\cdot}003 \\ \mu_\alpha \equiv y_0 = -0''{\cdot}300 \pm 0''{\cdot}008 \end{array}\right\} \text{(P.E.)}.$$

6·21. Equations of condition with more than three unknowns (Gauss's method)

The procedure of dealing with n equations of condition with more than three unknowns will be indicated in outline, taking the case of five unknowns which are denoted by x, y, z, u and v. The equations of condition, for which $n > 5$, are

$$\begin{array}{c|c} a_1 x + b_1 y + c_1 z + d_1 u + e_1 v = m_1 & s_1 \\ \cdots\cdots\cdots\cdots\cdots\cdots & \cdots \\ a_n x + b_n y + c_n z + d_n u + e_n v = m_n & s_n \end{array}$$

with the auxiliary quantities $s_1 \equiv a_1 + b_1 + c_1 + d_1 + e_1 + m_1$, etc., displayed on the right-hand side. The normal equations are formed in the usual way; they are

$$\begin{array}{c|c} x[aa] + y[ab] + z[ac] + u[ad] + v[ae] = [am] & [as] \\ x[ba] + y[bb] + z[bc] + u[bd] + v[be] = [bm] & [bs] \\ \cdots\cdots\cdots\cdots\cdots\cdots\cdots\cdots & \cdots \\ x[ea] + y[eb] + z[ec] + u[ed] + v[ee] = [em] & [es] \end{array}$$

with the usual checks

$$[aa] + [ab] + [ac] + [ad] + [ae] + [am] = [as], \quad \text{etc.}$$

For convenience the normal equations are written as

$$\mathrm{a}_1 x + \mathrm{a}_2 y + \mathrm{a}_3 z + \mathrm{a}_4 u + \mathrm{a}_5 v = M_1, \quad (1)$$

$$\mathrm{b}_1 x + \mathrm{b}_2 y + \mathrm{b}_3 z + \mathrm{b}_4 u + \mathrm{b}_5 v = M_2, \quad (2)$$

$$\mathrm{c}_1 x + \mathrm{c}_2 y + \mathrm{c}_3 z + \mathrm{c}_4 u + \mathrm{c}_5 v = M_3, \quad (3)$$

$$\mathrm{d}_1 x + \mathrm{d}_2 y + \mathrm{d}_3 z + \mathrm{d}_4 u + \mathrm{d}_5 v = M_4, \quad (4)$$

$$\mathrm{e}_1 x + \mathrm{e}_2 y + \mathrm{e}_3 z + \mathrm{e}_4 u + \mathrm{e}_5 v = M_5, \quad (5)$$

where $\mathrm{a}_1 \equiv [aa]$, $\mathrm{a}_2 \equiv [ab]$, etc. and $M_1 \equiv [am]$, etc.

We shall suppose for the moment that the normal equations have been solved, the values obtained being x_0, y_0, z_0, u_0 and v_0. As regards the precision of these solutions we require to find $[vv]$ and the weights $w_x, w_y, \ldots, w_v$ to be assigned to x_0, y_0, etc. The numerical value of $[vv]$ can be found either by calculating the values of the individual residuals, v_i, from the typical equation of condition by means of the formula

$$a_i x_0 + b_i y_0 + c_i z_0 + d_i u_0 + e_i v_0 - m_i = v_i,$$

or, alternatively, by means of the formula, given by 6·07 (4) and generalized for five unknowns,

$$[vv] = [mm] - M_1 x_0 - M_2 y_0 - M_3 z_0 - M_4 u_0 - M_5 v_0.$$

If μ is the standard error of the measures $m_1, m_2, \ldots, m_n$, then

$$\mu^2 = \frac{[vv]}{n-5}. \quad (6)$$

The standard error of x_0, for example, is then $\mu/\sqrt{w_x}$ for which w_x has to be evaluated by some process.

As the amount of calculation is considerable the solution of the normal equations is not to be undertaken lightly if economy in the numerical work of calculating the weights is to be kept in mind. There are two cases to be considered. In the first, the object of the investigation is the derivation of the most probable values of *all* the unknowns together with their standard or probable errors, the evaluation of each unknown being regarded as equally important from the point of view of the extension of precise knowledge relating to these quantities. In the second case, the object of the investigation consists primarily in deriving the values of, say, two unknowns with their standard or probable errors, although the equations of condition contain other unknowns whose values and precision are eventually of no particular interest; for example, in §§ 6·01 and 6·20 the equations of condition involve two unknowns Π ($\equiv x$) and μ_α ($\equiv y$) and it is the principal object of the investigation to determine the parallax, Π, of the star together with its measure of precision, the value of μ_α and its precision being of secondary interest.

We deal with the second case first, reflecting that it is essential, if the arithmetical work is to be reduced to a minimum, that the normal equations should be solved by some systematic procedure such as Gauss's. It is convenient to arrange the equations of condition and the normal equations in such a way that in the group (1)–(5), u and v are the two unknowns whose measures of precision are to be found; in Gauss's method we eliminate x, y and z successively, obtaining eventually two equations in u and v from which the values of u and v *and* their weights can be obtained with the minimum of effort. Thus we write (1) as

$$x = -\frac{\mathrm{b}_1}{\mathrm{a}_1}y - \frac{\mathrm{c}_1}{\mathrm{a}_1}z - \frac{\mathrm{d}_1}{\mathrm{a}_1}u - \frac{\mathrm{e}_1}{\mathrm{a}_1}v + \frac{M_1}{\mathrm{a}_1},$$

and substitute this expression for x in each of the remaining normal equations, calculating, for example, the 'unsimplified' quantity $\left(\mathrm{b}_2 - \frac{\mathrm{a}_2\mathrm{b}_1}{\mathrm{a}_1}\right)$ which is the coefficient of y in the second normal equation (2) when the above substitution of x is made. In this way four equations in y, z, u and v are obtained. The next step is to eliminate y by a similar procedure. The elimination of z then follows resulting in two equations of the forms:

$$Au + Bv = P, \tag{7}$$

$$Cu + Dv = Q, \tag{8}$$

in which A, B, ..., Q are 'unsimplified' quantities whose numerical values appear in the course of the calculations.

From (7),

$$u = -\frac{B}{A}v + \frac{P}{A},$$

and substitution in (8) gives

$$v\left(D - \frac{BC}{A}\right) = Q - \frac{PC}{A},$$

from which v is obtained; further, the coefficient of v is the weight w_v.

From (8),

$$v = -\frac{C}{D}u + \frac{Q}{D},$$

and substitution in (7) gives

$$u\left(A-\frac{BC}{D}\right)=P-\frac{BQ}{D};$$

u is then found and the coefficient of u in the last equation is the weight w_u. It must be emphasized that throughout the various steps in the eliminations the coefficients of the outstanding unknowns are *not* to be 'simplified' by the multiplication of any factor, however attractive such a proceeding may appear.

Once u and v and their weights have been obtained the values of the outstanding unknowns, x, y and z, can be found in any suitable way; their weights are not required, as the whole interest in the solution is deriving the values of u and v and their weights. The next step is to calculate $[vv]$ by either of the two methods mentioned earlier; then, by (6), μ is found. The results in which the investigator is really interested are:

$$u_0 \pm \frac{\mu}{\sqrt{w_u}}, \quad v_0 \pm \frac{\mu}{\sqrt{w_v}} \quad \text{(S.E.)}.$$

In terms of probable errors, the results are

$$u_0 \pm \frac{r}{\sqrt{w_u}}, \quad v_0 \pm \frac{r}{\sqrt{w_v}} \quad \text{(P.E.)},$$

where $r=0{\cdot}6745\mu$.

When *all* the values of the unknowns and their measures of precision are the concern of the investigator, the procedure is first to find u_0, v_0 and their weights by the elimination process just described. Then x_0, y_0 and w_x, w_y are found by the successive elimination of u, v and z. Finally, x_0, z_0 and w_x, w_z are obtained by the successive elimination of u, v and y, part of the work being the same as in the second series of elimination; the values of x_0 and w_x obtained in the second and third series should, of course, be the same. The work is checked throughout by the operations performed on the auxiliary quantities $s_1, s_2, \ldots, s_5$.

6·22. Other methods of solution

(i) The solution of the normal equations 6·21 (1)–(5) can clearly be effected in terms of determinants—a method† advocated in some quarters as being superior, in certain circumstances, to Gauss's method.

† E.g., E. T. Whittaker and G. Robinson, *The Calculus of Observations* (Blackie, 1924), p. 239, §121.

Consider the normal equations just referred to. Let Δ be the determinant defined by

$$\Delta = \begin{vmatrix} a_1, & a_2, & a_3, & a_4, & a_5 \\ b_1, & b_2, & b_3, & b_4, & b_5 \\ c_1, & c_2, & c_3, & c_4, & c_5 \\ d_1, & d_2, & d_3, & d_4, & d_5 \\ e_1, & e_2, & e_3, & e_4, & e_5 \end{vmatrix},$$

and let $A_1, A_2, \ldots, B_1, B_2, \ldots, E_1, E_2, \ldots,$ be the co-factors of $a_1, a_2, \ldots, b_1, b_2, \ldots, e_1, e_2$. Multiply 6·21 (1) by A_1, (2) by B_1, ...; (5) by E_1; then, by addition,

$$\begin{aligned} x(a_1A_1+b_1B_1+\ldots+e_1E_1)+y(a_2A_1+b_2B_1+\ldots+e_2E_1)+\ldots \\ +v(a_5A_1+b_5B_1+\ldots+e_5E_1) \\ =A_1M_1+B_1M_2+\ldots+E_1M_5. \end{aligned} \tag{1}$$

By the properties of determinants (see §6·09)

$$\left.\begin{aligned} a_1A_1+b_1B_1+\ldots+e_1E_1&=\Delta, \\ a_2A_1+b_2B_1+\ldots+e_2E_1&=0, \\ &\ldots\ldots \\ a_5A_1+b_5B_1+\ldots+e_5B_1&=0. \end{aligned}\right\} \tag{2}$$

Hence, if x_0 denotes the solution (1), then

$$\Delta x_0 = A_1M_1+B_1M_2+\ldots+E_1M_5. \tag{3}$$

Further, as in §6·18, the weight, w_x, of x_0 is given as $1/x'$, where x' is the corresponding solution of the normal equations in which M_1 is replaced by 1 and M_2, M_3, M_4 and M_5 are each replaced by 0. Thus, from (3), $\Delta x' = A_1$ so that

$$w_x = \frac{\Delta}{A_1}.$$

Here A_1 is the fourth-order determinant given by

$$A_1 = \begin{vmatrix} b_2, & b_3, & b_4, & b_5 \\ c_2, & \ldots, & \ldots, & c_5 \\ d_2, & \ldots, & \ldots, & d_5 \\ e_2, & \ldots, & \ldots, & e_5 \end{vmatrix}.$$

Now, a determinant of *any* order, say p, can be readily reduced to one of order $p-1$ and by successive applications of the procedure to one of third order and to one of the second order, the last being easily evaluated. To illustrate the procedure in a simple case, consider the determinant of third order, D, given by

$$D=\begin{vmatrix} x_1, & x_2, & x_3 \\ y_1, & y_2, & y_3 \\ z_1, & z_2, & z_3 \end{vmatrix} \equiv x_1 y_1 z_1 \begin{vmatrix} 1, & x_2/x_1, & x_3/x_1 \\ 1, & y_2/y_1, & y_3/y_1 \\ 1, & z_2/z_1, & z_3/z_1 \end{vmatrix}$$

or
$$D=x_1 y_1 z_1 \begin{vmatrix} 1, & x_2/x_1, & x_3/x_1 \\ 0, & y_2/y_1-x_2/x_1, & y_3/y_1-x_3/x_1 \\ 0, & z_2/z_1-x_2/x_1, & z_3/z_1-x_3/x_1 \end{vmatrix};$$

hence
$$D=\frac{1}{x_1}\begin{vmatrix} x_1 y_2-x_2 y_1, & x_1 y_3-x_3 y_1 \\ x_1 z_2-x_2 z_1, & x_1 z_3-x_3 z_1 \end{vmatrix}.$$

In the case of Δ, two successive applications of the above procedure reduce Δ to a third-order determinant and a further application to one of second order which can be readily evaluated.

In theory the method outlined is straightforward and in practice is comparatively easy of application when the coefficients a, b, ... in the normal equations are small positive or negative integers—for then the various determinants A_1, B_1, ... are readily reduced to calculable forms—or when only the evaluation of one or two of the unknowns (together with their standard errors) is regarded as the real object of the investigation. However, it is rarely the case that in actual problems the coefficients a, b, ... in the normal equations are small positive or negative integers and, accordingly, the method becomes one of considerable computational complexity; in such cases the systematic procedure of Gauss's method of solution is to be preferred.

One disadvantage of the determinant method is that such checks as are possible come only after a considerable amount of computation has been performed; if the checks fail there is nothing for it but to repeat the calculations up to this point.

(ii) A method involving matrix theory and describing the computation of determinants and the solution of linear equations has been given by T. Banachiewicz in *Acta Astronomica* (*Cracovie*), Ser. C, **3**, 41–72, to which further reference may be made by the reader.

6·23. Evaluation of the unknowns in equations of condition connected by rigorous equations

As a particular case consider n equations of condition in five unknowns x, y, z, u and v

$$a_i x+b_i y+c_i z+d_i u+e_i v=m_i \quad (i=1, 2, \ldots, n), \tag{1}$$

the unknowns being connected by two rigorous equations

$$\phi_1(x, y, z, u, v) = 0 \tag{2}$$

and
$$\phi_2(x, y, z, u, v) = 0. \tag{3}$$

We shall assume that the weight of (1) is w_i.

Now (2) and (3) are two equations which, theoretically, can be solved giving u and v each as functions of the three independent variables x, y and z. These solutions can be written symbolically as

$$u \equiv u(x, y, z), \quad v \equiv v(x, y, z). \tag{4}$$

When, in (1), u and v are eliminated by means of (4), the equations of condition are then expressed in terms of the three independent unknowns x, y and z.

If ϕ_1 and ϕ_2 are linear functions of x, y, ..., v, then the equations of condition are, after elimination of u and v, linear in x, y and z; the normal equations can then be formed in the usual way and their solution gives the most probable values of x, y and z; the most probable values of u and v are then obtained by means of (4).

Even when ϕ_1 and ϕ_2 are linear, the elimination of u and v from each of the n equations (1) is a cumbrous process. Instead, we consider the general problem when, at first, ϕ_1 and ϕ_2 are not necessarily linear.

Reduced to unit weight a typical equation of condition becomes

$$\sqrt{w_i}\,(a_i x + b_i y + \dots + e_i v) = \sqrt{w_i}\,.\,m_i.$$

If x, y, ..., v now denote the most probable values of the unknowns the sum of the squares of the residuals is

$$E \equiv \Sigma w_i (a_i x + \dots + d_i u + e_i v - m_i)^2. \tag{5}$$

If E' denotes this expression when $u \equiv u(x, y, z)$ and $v \equiv v(x, y, z)$ are substituted, E' is then a function of the three independent quantities x, y and z. The conditions for the sum of the squares of the residuals to be a minimum are

$$\frac{\partial E'}{\partial x} = \frac{\partial E'}{\partial y} = \frac{\partial E'}{\partial z} = 0.$$

Now
$$\frac{\partial E'}{\partial x} = \frac{\partial E}{\partial x} + \frac{\partial E}{\partial u}\frac{\partial u}{\partial x} + \frac{\partial E}{\partial v}\frac{\partial v}{\partial x} = 0 \tag{6}$$

by the first condition. There are two similar formulae in terms of y and z.

Similarly, from (2) and (3) when (4) are substituted, we have

$$\frac{\partial\phi_1}{\partial x}+\frac{\partial\phi_1}{\partial u}\frac{\partial u}{\partial x}+\frac{\partial\phi_1}{\partial v}\frac{\partial v}{\partial x}=0 \tag{7}$$

and

$$\frac{\partial\phi_2}{\partial x}+\frac{\partial\phi_2}{\partial u}\frac{\partial u}{\partial x}+\frac{\partial\phi_2}{\partial v}\frac{\partial v}{\partial x}=0. \tag{8}$$

Multiply (7) by λ_1 and (8) by λ_2, and add these results to (6). Then

$$\frac{\partial E}{\partial x}+\lambda_1\frac{\partial\phi_1}{\partial x}+\lambda_2\frac{\partial\phi_2}{\partial x}+\frac{\partial u}{\partial x}\left(\frac{\partial E}{\partial u}+\lambda_1\frac{\partial\phi_1}{\partial u}+\lambda_2\frac{\partial\phi_2}{\partial u}\right)$$
$$+\frac{\partial v}{\partial x}\left(\frac{\partial E}{\partial v}+\lambda_1\frac{\partial\phi_1}{\partial v}+\lambda_2\frac{\partial\phi_2}{\partial v}\right)=0. \tag{9}$$

There are two similar equations in terms of y and z, the coefficients of $\partial u/\partial y$ and $\partial v/\partial y$, of $\partial u/\partial z$ and $\partial v/\partial z$, being the same as in (9).

Suppose that λ_1 and λ_2 are such that the coefficients of $\partial u/\partial x$ and $\partial v/\partial x$ in (9) are zero; then we have the group of equations, five in number,

$$\left.\begin{aligned}
&\frac{\partial E}{\partial x}+\lambda_1\frac{\partial\phi_1}{\partial x}+\lambda_2\frac{\partial\phi_2}{\partial x}=0,\\
&\frac{\partial E}{\partial y}+\lambda_1\frac{\partial\phi_1}{\partial y}+\lambda_2\frac{\partial\phi_2}{\partial y}=0,\\
&\dots\dots\dots\dots\dots\dots\dots\dots\\
&\frac{\partial E}{\partial v}+\lambda_1\frac{\partial\phi_1}{\partial v}+\lambda_2\frac{\partial\phi_2}{\partial v}=0.
\end{aligned}\right\} \tag{10}$$

These, together with the two equations from (2) and (3), namely, $\phi_1=0$ and $\phi_2=0$, are *seven* equations from which the *seven* unknowns x, y, z, u, v and λ_1, λ_2 can be found. In (10), $\partial E/\partial x$, etc., are found at once from (5); thus, for example,

$$\frac{1}{2}\frac{\partial E}{\partial x}=x[waa]+y[wab]+\dots+v[wae]-[wam].$$

It is clear that the procedure can be generalized for k unknowns connected by p $(p<k)$ rigorous equations; the number of *independent* unknowns is $k-p$, and for the principle of least squares to be applicable the number, n, of equations of conditions must be greater than $k-p$.

The simplest case occurs when each of the functions ϕ is a *linear* function of the unknowns; then the equations (10) are linear in x, y, $\dots$, v and in λ_1, λ_2; the elimination of λ_1 and λ_2 leaves five linear equations in the five unknowns from which the values of the unknowns are obtained. The simple problem in § 6·24, involving three unknowns and one rigorous linear equation, illustrates the general method of solution, together with an assessment of the precision of the unknowns.

When at least one of the ϕ's is *non-linear* the general method involves solutions of complicated equations; the difficulties are avoided if approximate values of the unknowns can be obtained by any means, for then each ϕ can be reduced to a linear equation by the application of Taylor's theorem for the expansion of a function. The procedure is illustrated in § 6·24 (ii).

6·24. Illustrative examples

(i). *To find the most probable values of the angles x, y and z of a triangle from measures α, β and γ respectively.*

In this simple problem the equations of condition are

$$x-\alpha=0, \quad y-\beta=0, \quad z-\gamma=0$$

with the Euclidean condition

$$\phi \equiv x+y+z-\pi=0. \tag{1}$$

If the weights of α, β and γ are w_1, w_2 and w_3 then, from 6·23 (10), the most probable values of the unknowns are given by

$$\frac{\partial E}{\partial x}+\lambda_1\frac{\partial \phi}{\partial x}=0 \tag{2}$$

with two similar equations in y and z, where

$$E \equiv w_1(x-\alpha)^2+w_2(y-\beta)^2+w_3(z-\gamma)^2. \tag{3}$$

Now, from (1), $$\frac{\partial \phi}{\partial x}=\frac{\partial \phi}{\partial y}=\frac{\partial \phi}{\partial z}=1;$$

hence (2) becomes $$2w_1(x-\alpha)+\lambda_1=0,$$

or, on writing $\lambda_1=-2\lambda$ for simplicity,

$$x=\alpha+\frac{\lambda}{w_1}. \tag{4}$$

Similarly, $$y=\beta+\frac{\lambda}{w_2}, \quad z=\gamma+\frac{\lambda}{w_3}.$$

Substitute these in (1); then

$$\alpha+\beta+\gamma+\lambda\left(\frac{1}{w_1}+\frac{1}{w_2}+\frac{1}{w_3}\right)=\pi,$$

from which, on writing $U=w_1w_2+w_2w_3+w_3w_1$,

$$\lambda=\{\pi-(\alpha+\beta+\gamma)\}\frac{w_1w_2w_3}{U}.$$

Substitution of this expression in (4) gives

$$x=\frac{w_1(w_2+w_3)}{U}\alpha-\frac{w_2w_3}{U}(\beta+\gamma-\pi). \tag{5}$$

This expression for x gives the most probable value of the angle whose measure is α.

If the measure α is the result of a series of individual measures of unit weight so that r_α is the probable error of α, the probable error of measures of unit weight being r, then

$$r_\alpha^2=\frac{r^2}{w_1}.$$

Similarly, the probable errors r_β and r_γ of the measures β and γ are given by

$$r_\beta^2=\frac{r^2}{w_2},\quad r_\gamma^2=\frac{r^2}{w_3}.$$

From (5), on applying the general theorem of § 4·14, the probable error of x, denoted by r_x, is given by

$$r_x^2=\frac{w_1^2}{U^2}(w_2+w_3)^2r_\alpha^2+\frac{w_2^2w_3^2}{U^2}(r_\beta^2+r_\gamma^2),$$

which, on reduction becomes,

$$r_x^2=\frac{(w_2+w_3)\,r^2}{U}. \tag{6}$$

The probable errors of y and z are given by analogous formulae.

It is to be remarked that in this simple example E can be expressed, by means of (1), in terms of two independent variables; thus we have, by eliminating z between (1) and (3),

$$E'=w_1(x-\alpha)^2+w_2(y-\beta)^2+w_3(x+y+\gamma-\pi)^2,$$

and the conditions, $\partial E'/\partial x=0$ and $\partial E'/\partial y=0$, for the most probable values of x and y become

$$w_1(x-\alpha)+w_3(x+y+\gamma-\pi)=0,\quad w_2(y-\beta)+w_3(x+y+\gamma-\pi)=0$$

or
$$(w_1+w_3)\,x+w_3y=w_1\alpha+w_3(\pi-\gamma)$$

and
$$w_3x+(w_2+w_3)\,y=w_2\beta+w_3(\pi-\gamma).$$

From these we obtain the most probable value of x, namely,

$$x=\frac{w_1(w_2+w_3)}{U}\alpha-\frac{w_2w_3}{U}(\beta+\gamma-\pi),$$

as in (5). The formula for r_x in (6) then follows as previously shown.

(ii). *To find the most probable position of a ship derived from n position lines, subject to a given condition.*

As a practical problem we suppose that the position lines are obtained from compass bearings of n shore objects and that the horizontal sextant angle (H.S.A.) of two suitably chosen objects, A and B, has been measured. A compass bearing can be read, we shall suppose, to $\frac{1}{2}°$ and the H.S.A. to $0'{\cdot}1$; if we regard the latter as accurate in comparison with the former, the H.S.A. places the ship on a *particular circle* passing through A and B, the angle subtended at the ship by the chord AB being the value of the H.S.A.

If the centre of the circle is taken as the origin of coordinates, together with suitably oriented axes, the equation of the circle is

$$\phi(X,Y)\equiv X^2+Y^2-K^2=0. \tag{1}$$

Referred to this coordinate system, the equation of a typical position line can be written as

$$a_iX+b_iY-M_i=0, \tag{2}$$

in which M_i denotes a measured quantity with known associated values of a_i and b_i; also, the weight is w_i.

Applying the general theory in §6·23 we form the function E given by

$$E=\Sigma w_i(a_iX+b_iY-M_i)^2.$$

For the most probable position of the ship, subject to the condition (1), we have

$$\frac{\partial E}{\partial X}+\lambda\frac{\partial \phi}{\partial X}=0,\quad \frac{\partial E}{\partial Y}+\lambda\frac{\partial \phi}{\partial Y}=0.$$

These reduce to two linear equations of the forms

$$pX+qY-s+\lambda X=0,\quad qX+rY-t+\lambda Y=0$$

from which

$$X=\frac{s(\lambda+r)-qt}{(\lambda+p)(\lambda+r)-q^2},\quad Y=\frac{t(\lambda+p)-qs}{(\lambda+p)(\lambda+r)-q^2}.$$

Substitution of these in (1) gives

$$\{s(\lambda+r)-qt^2\}+\{t(\lambda+p)-qs\}^2=K^2\{(\lambda+p)(\lambda+r)-q^2\}^2,$$

which is a quartic equation in λ.

The solution of this equation would, in general, involve considerable computation, together with a discussion as to which of the four roots is the relevant one.

A much simpler procedure than that just described is to find approximate coordinates (X_0, Y_0) which can be taken as the solution values of any two equations of condition with widely different numerical values of the gradients of the lines representing the equations of condition. We can then write

$$X = X_0 + x, \quad Y = Y_0 + y,$$

in which x and y are small quantities whose squares and products can be neglected. A typical equation of condition, (2), can then be written as

$$a_i x + b_i y - m_i = 0,$$

where

$$m_i = M_i - a_i X_0 - b_i Y_0.$$

In the same way (1) becomes

$$\phi \equiv \phi(X_0, Y_0) + 2X_0 x + 2Y_0 y - K^2 = 0$$

or

$$\phi \equiv 2X_0 x + 2Y_0 y - k^2 = 0, \tag{3}$$

where

$$k^2 = K^2 - (X_0^2 + Y_0^2).$$

The conditions for the most probable values of x and y are then given by

$$\frac{\partial}{\partial x} \Sigma w_i (a_i x + b_i y - m_i)^2 + \lambda \frac{\partial}{\partial x} (2X_0 x + 2Y_0 y - k^2) = 0,$$

together with a similar equation in y. These are two linear equations in x, y *and* λ of the forms

$$px + qy - u + \lambda X_0 = 0, \quad qx + ry - v + \lambda Y_0 = 0,$$

from which x and y can be found as linear functions of λ of the forms

$$x = A\lambda + B \quad \text{and} \quad y = C\lambda + D. \tag{4}$$

Substitution of these in (3) yields the value of λ which, entered in (4), gives the most probable values of x and y, and hence of X and Y.

6·25. Precision of a function of several unknowns determined from *n* equations of condition

We consider a function $\phi(x, y, z)$ of three variables x, y and z which are evaluated, *in an independent investigation*, from n equations of condition (all reduced to unit weight) of the form

$$a_i x + b_i y + c_i z - m_i = 0, \tag{1}$$

in which, as usual, m_i corresponds to a measured quantity with associated error ϵ_i. From (1) the normal equations are

$$\left.\begin{aligned} \mathrm{a}x+\mathrm{h}y+\mathrm{g}z=M_1,\\ \mathrm{h}x+\mathrm{b}y+\mathrm{f}z=M_2,\\ \mathrm{g}x+\mathrm{f}y+\mathrm{c}z=M_3,\end{aligned}\right\} \tag{2}$$

where, in the notation of § 6·09, $\mathrm{a}=[aa]$, $\mathrm{h}=[ab]$, etc., and $M_1=[am]$, etc. The solutions of (2) are x_0, y_0 and z_0, the standard errors of which are μ_x, μ_y and μ_z.

Our object is to find the precision of $\phi(x_0, y_0, z_0)$.

Now, from 6·12 (3) and (4), the *true* values of x, y and z are given in term of the errors ϵ_i by

$$x-x_0=[\alpha\epsilon], \quad y-y_0=[\beta\epsilon] \quad \text{and} \quad z-z_0=[\gamma\epsilon], \tag{3}$$

where α, β and γ are defined by 6·09 (16), (17) and (18).

For convenience we take the error, e, of $\phi(x_0, y_0, z_0)$ in the sense

$$e=\phi(x, y, z)-\phi(x_0, y_0, z_0); \tag{4}$$

then, assuming that the errors ϵ_i in (3) are small, we can write (4), with sufficient accuracy by means of Taylor's expansion, as

$$e=[\alpha\epsilon]\left(\frac{\partial\phi}{\partial x}\right)_0+[\beta\epsilon]\left(\frac{\partial\phi}{\partial y}\right)_0+[\gamma\epsilon]\left(\frac{\partial\phi}{\partial z}\right)_0$$

or

$$e=[\alpha\epsilon]L+[\beta\epsilon]M+[\gamma\epsilon]N,$$

where

$$L=\left(\frac{\partial\phi}{\partial x}\right)_0, \quad M=\left(\frac{\partial\phi}{\partial y}\right)_0, \quad N=\left(\frac{\partial\phi}{\partial z}\right)_0. \tag{5}$$

Thus,

$$e=\Sigma(\alpha_i L+\beta_i M+\gamma_i N)\,\epsilon_i. \tag{6}$$

Let μ be the standard error of the errors ϵ_i; then, by § 6·12,

$$\mu^2=\frac{[vv]}{n-3},$$

where v_i is the residual when x_0, y_0 and z_0 are substituted in (1).

Now, in (6), e is a linear function of the errors, ϵ_i; hence, if μ_ϕ denotes the standard error of e, that is, of the function ϕ, then, by § 4·14,

$$\begin{aligned}\mu_\phi^2&=\mu^2\Sigma(\alpha_i L+\beta_i M+\gamma_i N)^2\\ &=\mu^2[[\alpha\alpha]L^2+[\beta\beta]M^2+[\gamma\gamma]N^2+2[\beta\gamma]MN\\ &\qquad+2[\gamma\alpha]NL+2[\alpha\beta]LM],\end{aligned}$$

or, by 6·11 (3) and (4),

$$\mu_\phi^2 = \frac{\mu^2}{\Delta}[AL^2 + BM^2 + CN^2 + 2FMN + 2GNL + 2HLM], \qquad (7)$$

where Δ is the determinant associated with the normal equations and A, B, ... are the usual co-factors of the determinant. But, from 6·13 (3) and (4),

$$\mu_x^2 = \frac{A}{\Delta}\mu^2, \quad \mu_y^2 = \frac{B}{\Delta}\mu^2, \quad \mu_z^2 = \frac{C}{\Delta}\mu^2;$$

hence, (7) becomes

$$\mu_\phi^2 = L^2\mu_x^2 + M^2\mu_y^2 + N^2\mu_z^2 + \frac{2\mu^2}{\Delta}(FMN + GNL + HLM), \qquad (8)$$

in which L, M and N are to be replaced by the expressions in (5).

This formula (9) gives the standard error of the function ϕ computed in terms of the solution values x_0, y_0 and z_0 of the normal equations (2) and the corresponding standard errors; it can be readily transformed into a formula in terms of probable errors. Thus, if r is the probable error of the errors ϵ_i, that is, of the measures m_i, and r_x, r_y and r_z are the probable errors of x_0, y_0 and z_0, then

$$r_\phi^2 = L^2r_x^2 + M^2r_y^2 + N^2r_z^2 + \frac{2r^2}{\Delta}(FMN + GNL + HLM). \qquad (9)$$

6·26. Precision of the coordinates of the solar apex and of the solar velocity derived from measures of radial velocity

We illustrate the principles of the preceding section by considering a standard problem in stellar kinematics.

If $(-x, -y, -z)$ are the equatorial components of the sun's linear velocity, the measures of the radial velocities of the stars in a small region of the sky furnish an equation of condition of the form

$$a_ix + b_iy + c_iz - m_i = 0, \qquad (1)$$

where m_i is the algebraic mean of the observed radial velocities of the stars in this region and a_i, b_i and c_i are known functions of the coordinates of the centre of the region. In general, the weight to be attached to (1) is n_i, the number of stars observed in the region; however, it will be assumed that such an equation of condition is reduced to one of unit weight and thus takes the form of (1).

As in the previous section, x_0, y_0 and z_0 are the solutions of the normal equations and μ is the standard error associated with the measures; also μ_x, μ_y and μ_z are the standard errors of x_0, y_0 and z_0.

The right ascension, α_0, and the declination, δ_0, of the solar apex are given,† with an obvious change in notation, by

$$\alpha_0 = \tan^{-1}(y/x), \quad \delta_0 = \tan^{-1}\{z/\sqrt{(x^2+y^2)}\}$$

and the solar velocity, v, is given by

$$v = \sqrt{(x^2+y^2+z^2)}.$$

(i) *Precision of* α_0. In this case, $\phi = \tan^{-1}(y/x)$, from which

$$\frac{\partial\phi}{\partial x} = -\frac{y}{x^2+y^2}, \quad \frac{\partial\phi}{\partial y} = \frac{x}{x^2+y^2}, \quad \frac{\partial\phi}{\partial z} = 0.$$

It is found that α_0 is very close to 270° so that, effectively, x_0/y_0 can be taken to be zero. Thus

$$L \equiv \left(\frac{\partial\phi}{\partial x}\right)_0 = -\frac{1}{y_0}, \quad M = N = 0$$

If μ_α denotes the standard error of α_0 then, by 6·25 (8),

$$\mu_\alpha = \frac{\mu_x}{y_0}.$$

In terms of probable error, since $r = 0{\cdot}6745\mu$,

$$r_\alpha = \frac{r_x}{y_0}.$$

(ii) *Precision of* δ_0. In this case, $\phi = \tan^{-1}\{z/\sqrt{(x^2+y^2)}\}$, from which

$$\frac{\partial\phi}{\partial x} = -\frac{xz}{v^2\sqrt{(x^2+y^2)}}, \quad \frac{\partial\phi}{\partial y} = -\frac{yz}{v^2\sqrt{(x^2+y^2)}}, \quad \frac{\partial\phi}{\partial z} = \frac{1}{v^2}\sqrt{(x^2+y^2)}.$$

With the approximation introduced in (i), we have

$$L = 0, \quad M = -\frac{z_0}{v_0^2}, \quad N = \frac{y_0}{v_0^2}.$$

Hence, from 6·25 (8),

$$\mu_\delta^2 = \frac{1}{v_0^4}\left(z_0^2\mu_y^2 + y_0^2\mu_z^2 - \frac{2\mu^2}{\Delta}Fy_0z_0\right). \tag{1}$$

The formula for r_δ can be derived at once by means of 6·25 (9).

(iii) *Precision of* v_0. In this case $\phi = \sqrt{(x^2+y^2+z^2)}$, from which

$$\frac{\partial\phi}{\partial x} = \frac{x}{v}, \quad \frac{\partial\phi}{\partial y} = \frac{y}{v}, \quad \frac{\partial\phi}{\partial z} = \frac{z}{v}.$$

† W. M. Smart, *Spherical Astronomy*, 4th ed. (Cambridge University Press, 1956), pp. 273, 275.

Then, $L=0$, $M=y_0/v_0$, $N=z_0/v_0$ and, by 6·26 (8),

$$\mu_v^2=\frac{1}{v_0^2}\left(y_0^2\mu_y^2+z_0^2\mu_z^2+\frac{2\mu^2}{\Delta}Fy_0z_0\right). \tag{2}$$

The formula for r_v follows.

It is to be remarked that, if μ_δ has been calculated, the value of μ_v can be found very simply; by the elimination of F between (1) and (2) we have, with the approximation introduced,

$$\mu_v^2+v_0^2\mu_\delta^2=\mu_y^2+\mu_z^2.$$

Also

$$r_v^2+v_0^2r_\delta^2=r_y^2+r_z^2.$$

CHAPTER 7

THEORETICAL FREQUENCY DISTRIBUTIONS

7·01. Calculated and theoretical moments

In this chapter we consider the problem of representing a statistical distribution of a variable x by a function of x. More precisely, if y_i is the frequency in the class interval bounded by $x_i - \frac{1}{2}c$ and $x_i + \frac{1}{2}c$, we shall obtain the form of the function $F_1(x)$ which best represents the series of points (x_i, y_i) or, alternatively, the form of the function $F(x)$ which best represents the *relative* frequency distribution, that is, the series of points $(x_i, y_i/N)$, where N is the total frequency. The next section deals with the former; thereafter, the theoretical treatment is mainly concerned with the relative frequency distribution.

It will be assumed that the moments of various orders have been calculated, from the statistics according to the processes described in Chapter 1. The *calculated* moment of order r will be denoted by m_r with reference to the mean $\bar{x}$ and by $m_r(0)$ with reference to the origin. Thus

$$m_r = \frac{1}{N}\Sigma(x_i - \bar{x})^r y_i$$

and

$$m_r(0) = \frac{1}{N}\Sigma x_i^r y_i \equiv \frac{1}{N}[x^r y],$$

the last expression being in the Gaussian notation.

For the *theoretical* relative distribution given by $y = F(x)$, the moments of order r with respect to the mean and to the origin are denoted by μ_r and $\mu_r(0)$ respectively; then

$$\mu_r = \int (x - \bar{x})^r F(x)\,dx = \int (x - \bar{x})^r y\,dx$$

and

$$\mu_r(0) = \int x^r y\,dx,$$

the integration being performed between the appropriate limits of x furnished by the data.

The characteristics of the function $F(x)$—and of $F_1(x)$—will then be deduced by identifying the theoretical moments with the known calculated moments.

7·02. Representation of a frequency distribution by a polynomial

When we are given the frequency y_i (>0) corresponding to the value x_i of the variable x the limits of which are x_1 and x_2 $(x_2 > x_1)$, the distribution shown by plotting the points (x_i, y_i) as in Fig. 13 may suggest a polynomial representation. For example, the curve may be a section of the parabola with the equation

$$y = A + Bx + Cx^2. \tag{1}$$

If we write $x_2 \equiv x_1 + k$, we can conveniently suppose that (1) is referred to $O_1(x_1, 0)$ as origin, the variable now lying in the range $0 \leqslant x \leqslant k$.

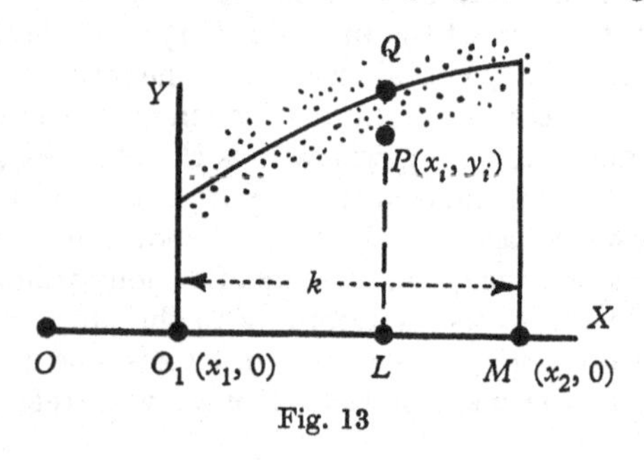

Fig. 13

If $P(x_i, y_i)$ is a statistical point and Q the point on (1) with abscissa x_i the difference, PQ, leads to the equation of condition

$$A + Bx_i + Cx_i^2 - y_i = 0,$$

in which A, B and C are unknowns; the most probable values of these are found from the normal equations

$$\left.\begin{aligned} A + B\Sigma x_i + C\Sigma x_i^2 &= \Sigma y_i = N, \\ A\Sigma x_i + B\Sigma x_i^2 + C\Sigma x_i^3 &= \Sigma x_i y_i = Nm_1(0), \\ A\Sigma x_i^2 + B\Sigma x_i^3 + C\Sigma x_i^4 &= \Sigma x_i^2 y_i = Nm_2(0), \end{aligned}\right\} \tag{2}$$

the numerical values of $m_1(0)$ and $m_2(0)$ being supposed known; the precision, if required, of each of A, B and C can then be obtained by means of the methods described in Chapter 6.

The theoretical moments $\mu_r(0)$ are given, for $r = 0$, 1 and 2, by

$$N = \int_0^k y\,dx = \int_0^k (A + Bx + Cx^2)\,dx,$$

$$N\mu_1(0) = \int_0^k xy\,dx \quad \text{and} \quad N\mu_2(0) = \int_0^k x^2 y\,dx$$

from which it is easily found that

$$A+\tfrac{1}{2}Bk+\tfrac{1}{3}Ck^2=1/(Nk),$$

$$\tfrac{1}{2}A+\tfrac{1}{3}Bk+\tfrac{1}{4}Ck^2=\mu_1(0)/(Nk^2),$$

$$\tfrac{1}{3}A+\tfrac{1}{4}Bk+\tfrac{1}{5}Ck^2=\mu_2(0)/(Nk^3).$$

If, in these, $\mu_1(0)$ and $\mu_2(0)$ are identified with the calculated moments $m_1(0)$ and $m_2(0)$, the solution of the equations gives values of A, B and C which should be equal to, or in close approximation with, the values derived from the solution of the equations (2).

If a closer approximation is to be found, the equation of the theoretical curve can be taken to be

$$y=A+Bx+Cx^2+Dx^3,$$

in which case the moment of highest order to be calculated is $m_3(0)$.

7·03. Representation of a frequency distribution by a trigonometrical series

Subject to certain general conditions which some theoretical statistical distributions satisfy, a function of x can be represented by the Fourier expansion

$$y\equiv f(x)=A_0+\sum_{k=1}^{\infty}(A_k\cos kx+B_k\sin kx),$$

where $k=1, 2, \ldots$. This form of $f(x)$ is particularly relevant when the statistics possess the characteristics of periodicity—as in the case of the heights of the tide during a lunar month. In practice, it is sufficient to assume that $f(x)$ is represented adequately by a finite number of terms.

Let x_0 $(\equiv 0)$ and x_n be the ends of the range and suppose that the range is divided at $x_1, x_2, \ldots, x_{n-1}$ into n equal sections. Choose the unit of x to be $2\pi/n$; then

$$x_r=\frac{2\pi r}{n}.$$

We consider the frequencies, y_r, for values of $r=1, 2, \ldots, n$; then

$$y_r=A_0+\sum_k\left(A_k\cos\frac{2\pi kr}{n}+B_k\sin\frac{2\pi kr}{n}\right), \tag{1}$$

where k is now limited. We thus have $(2k+1)$ constants and n values of the frequency. If $k=2n+1$ the number of equations (1) is sufficient for the calculation of the constants A and B.

Generally, however, the number of terms in (1) is smaller than n; for example, if y is the mean monthly temperature throughout the year ($n=12$), the statistics may be adequately represented by seven terms in (1), corresponding to the values $k=1$, 2 and 3. The constants are then evaluated by a least-squares solution.

Assuming that $2k+1<n$, we have the normal equation corresponding to A_0 given by

$$\sum_{r=1}^{n} y_r = nA_0 + \sum_k \left[A_k \sum_{r=1}^{n} \cos\frac{2\pi kr}{n} + B_k \sum_{r=1}^{n} \sin\frac{2\pi kr}{n} \right]. \tag{2}$$

Now
$$\sum_{r=1}^{n} \cos r\alpha = \frac{\sin\frac{1}{2}n\alpha}{\sin\frac{1}{2}\alpha} \cos\tfrac{1}{2}(n+1)\alpha \tag{3}$$

and
$$\sum_{r=1}^{n} \sin r\alpha = \frac{\sin\frac{1}{2}n\alpha}{\sin\frac{1}{2}\alpha} \sin\tfrac{1}{2}(n+1)\alpha. \tag{4}$$

Hence, if $\alpha=2\pi k/n$, then $\sin\frac{1}{2}n\alpha \equiv \sin k\pi = 0$; consequently, each summation for r in (2) vanishes so that

$$A_0 = \frac{1}{n}\Sigma y_r. \tag{5}$$

The normal equation corresponding to a *particular* A_k is obtained by multiplying (1) by $\cos 2\pi kr/n$ and summing for r. Then, if $j \neq k$,

$$\begin{aligned}\sum_r y_r \cos\frac{2\pi kr}{n} &= A_0 \sum_r \cos\frac{2\pi kr}{n} + A_k \sum_r \cos^2\frac{2\pi kr}{n} \\ &\quad + \tfrac{1}{2}B_k \sum_r \sin\frac{4\pi kr}{n} + \sum_j A_j \sum_r \cos\frac{2\pi kr}{n}\cos\frac{2\pi jr}{n} \\ &\quad + \sum_j B_j \sum_r \cos\frac{2\pi kr}{n}\sin\frac{2\pi jr}{n}.\end{aligned}$$

By (3), the coefficient of A_0 is zero. The coefficient of A_k is

$$\tfrac{1}{2}n + \tfrac{1}{2}\Sigma\cos\frac{4\pi kr}{n} \quad \text{or} \quad \tfrac{1}{2}n,$$

by (3). The coefficient of A_j is

$$\tfrac{1}{2}\sum_r \left\{ \cos\frac{2\pi(k+j)r}{n} + \cos\frac{2\pi(k-j)r}{n} \right\}$$

and each summation vanishes by (3). Similarly, the coefficients of B_k and B_j vanish. Hence

$$A_k = \frac{2}{n}\sum_r y_r \cos\frac{2\pi kr}{n}. \tag{6}$$

Similarly,
$$B_k = \frac{2}{n}\sum_r y_r \sin\frac{2\pi kr}{n}. \tag{7}$$

7·04. Precision of the constants

When the constants A and B have been found by means of 7·03 (5), (6) and (7), the residual v_r corresponding to y_r is given by

$$v_r = y_r - A_0 - \sum_k \left(A_k \cos \frac{2\pi kr}{n} + B_k \sin \frac{2\pi kr}{n} \right). \tag{1}$$

Then

$$\sum v_r = \sum y_r - nA_0 - \sum_k \left[A_k \sum_r \cos \frac{2\pi kr}{n} + B_k \sum \sin \frac{2\pi kr}{n} \right].$$

Hence, by 7·03 (3), (4) and (5), $\Sigma v_r = 0$.

The derivation of the normal equations in the previous section for the A's and B's, regarded as unknowns, is equivalent to the application of the principle of least squares, namely, that $V \equiv \Sigma v_r^2$ is a minimum with respect to the A's and B's; then, for a particular A_k,

$$\frac{\partial V}{\partial A_k} \equiv 2\Sigma v_r \frac{\partial v_r}{\partial A_k} = 0,$$

or, from (1),

$$\sum_r v_r \cos \frac{2\pi kr}{n} = 0. \tag{2}$$

Similarly,

$$\sum_r v_r \sin \frac{2\pi kr}{n} = 0. \tag{3}$$

Now, by (1),

$$V = \sum_r v_r \left[y_r - A_0 - \sum_k \left(A_k \cos \frac{2\pi kr}{n} + B_k \sin \frac{2\pi kr}{n} \right) \right],$$

or, by (2) and (3) and since $\Sigma v_r = 0$,

$$V = \Sigma v_r y_r;$$

hence, by (1),

$$[vv] = [yy] - A_0 \Sigma y_r - \sum_k \left[A_k \sum_r y_r \cos \frac{2\pi kr}{n} + B_k \sum_r y_r \sin \frac{2\pi kr}{n} \right],$$

and, by 7·03 (5), (6) and (7),

$$[vv] = [yy] - nA_0^2 - \frac{n}{2} \Sigma (A_k^2 + B_k^2), \tag{4}$$

from which the numerical value of $[vv]$ is readily derived.

If μ denotes the standard deviation of the quantities y_r then, by 6·12 (11),

$$\mu^2 = \frac{[vv]}{n - (2k+1)}, \tag{5}$$

since $2k+1$ is the number of unknowns.

In 7·03 (6), A_k is a linear function of the y's; hence, if ν denotes the standard error of A_k, then, by 3·07 (11),

$$\nu^2 = \frac{4\mu^2}{n^2} \sum_r \cos^2 \frac{2\pi kr}{n} = \frac{2\mu^2}{n}.$$

Thus, ν is the same for all the A_k's and, as it is easily seen, for all the B_k's.

Also, if ν_0 denotes the standard error of A_0, then, from 7·03 (5),

$$\nu_0^2 = \frac{\mu^2}{n}.$$

The solutions, with their precision expressed in terms of standard error, are:

$$A_0 = \frac{1}{n} \Sigma y_r \pm \frac{\mu}{\sqrt{n}},$$

$$A_k = \frac{2}{n} \Sigma y_r \cos \frac{2\pi kr}{n} \pm \mu \sqrt{\left(\frac{2}{n}\right)},$$

$$B_k = \frac{2}{n} \Sigma y_r \sin \frac{2\pi kr}{n} \pm \mu \sqrt{\left(\frac{2}{n}\right)},$$

where μ is given by means of (4) and (5).

The precision in terms of probable error can be expressed in the usual way.

7·05. The Gram-Charlier series

This is a series of terms which takes into account the departure of a univariate distribution from a normal distribution. The series was derived on theoretical grounds by Charlier† on the basis of the conception of 'elementary errors', discussed in § 2·18 in connexion with Hagen's derivation of the normal function.

Denote the variable by x; then the statistics furnish (i) the mean $\bar{x}$, and (ii) the standard deviation, σ, together with the higher moments.

Let $f(x)$ denote the normal function with the parameters $\bar{x}$ and σ; then

$$f(x) = \frac{1}{\sigma\sqrt{(2\pi)}} e^{-(x-\bar{x})^2/2\sigma^2}. \tag{1}$$

We use this form of the normal function as being more convenient when we consider Hermite's polynomials later in connexion with the Gram-Charlier series.

† C. V. L. Charlier, *Arkiv. för Matematik, Astronomi och Fysik.* Bd. 2, no. 8, 1905; see also *Medd. från Lunds Astron. Observatorium*, ser. II, no. 4, 1906.

If $F(x)$ is the relative distribution function associated with the statistics, taken over the complete range of values of x from $-\infty$ to ∞, the Gram-Charlier series is written

$$F(x)=\sum_{n=0}\frac{A_n}{n!}\frac{d^nf(x)}{dx^n},\tag{2}$$

in which the A's are constants to be determined eventually in terms of the principal moments derived from the statistics. The series in (2) is sometimes referred to as the *A-series*.

By definition,

$$\int_{-\infty}^{\infty}F(x)\,dx=1.\tag{3}$$

Also, the theoretical moments, μ_r, about the mean, $\bar{x}$, are given by

$$\mu_r=\int_{-\infty}^{\infty}(x-\bar{x})^r F(x)\,dx.\tag{4}$$

The use of a series such as (2) implies, on general grounds, the necessity for a discussion of the convergency of the series, which may not be immediately practicable. This difficulty can be overcome by regarding the statistical distribution to be adequately represented by a certain number of terms—say, up to that with $n=6$; in such cases it is presumed that the statistical distribution is closely allied to a normal distribution and, accordingly, the formula for $F(x)$ is then essentially an empirical formula.

In the calculation of the several constants, A, the *theoretical* principal moments, μ_r, will be equated to the corresponding *calculated* moments, m_r, derived according to the principles described in Chapter 1. In particular, since x is the mean, $m_1=0$ and, identifying m_1 with μ_1, we have

$$\mu_1=0.\tag{5}$$

Similarly,

$$\mu_2=\sigma^2.\tag{6}$$

It is to be remembered that $\bar{x}$ and σ are now supposed to be known.

7·06. Transformation of the Gram-Charlier series

In 7·05 (1), write

$$t=(x-\bar{x})/\sigma,\tag{1}$$

and let $\phi(t)$ be defined by

$$\phi(t)=\frac{1}{\sqrt{(2\pi)}}e^{-\frac{1}{2}t^2},\tag{2}$$

with the obvious property, on writing $t=\tau\sqrt{2}$,

$$\int_{-\infty}^{\infty}\phi(t)\,dt\equiv\frac{1}{\sqrt{\pi}}\int_{-\infty}^{\infty}e^{-\tau^2}d\tau=1.\tag{3}$$

Then, from 7·05 (1), $$f(x)=\frac{1}{\sigma}\phi(t). \tag{4}$$

Also, from (1), $$\frac{d}{dx}=\frac{1}{\sigma}\frac{d}{dt}=\frac{1}{\sigma}D,$$

where D denotes the operator d/dt; further,

$$\frac{d^n}{dx^n}=\frac{1}{\sigma^n}D^n,$$

so that $$\frac{d^nf(x)}{dx^n}=\frac{1}{\sigma^{n+1}}D^n\phi. \tag{5}$$

Express $F(x)$ in terms of t by means of (1), and let

$$\sigma F(x)=\psi(t); \tag{6}$$

then $$\psi(t)=\sum_{n=0}\frac{A_n}{\sigma^n n!}D^n\phi,$$

or, if we write $$a_n=\frac{A_n}{\sigma^n n!}, \tag{7}$$

then $$\psi(t)=\sum_0 a_n D^n\phi. \tag{8}$$

From 7·05 (3), we have at once, by (1) and (6),

$$\int_{-\infty}^{\infty}\psi(t)\,dt=1. \tag{9}$$

Also, from 7·05 (4), $$\mu_r=\sigma^r\int_{-\infty}^{\infty}t^r\psi(t)\,dt. \tag{10}$$

7·07. Hermite's polynomials

Hermite's polynomials, $H_n(t)$, are defined in terms of the function $\phi(t)$ by

$$H_n(t)\,\phi(t)=(-1)^n D^n\phi. \tag{1}$$

The first few polynomials are readily found; thus $H_0=1$, and since, from 7·06 (2),

$$D\phi=-t\phi, \tag{2}$$

then $H_1=t$; also, $D^2\phi\equiv-D(t\phi)=(t^2-1)\,\phi$, so that $H_2=t^2-1$.

A general formula is obtained by applying Leibniz's rule to (2); thus

$$D^{n+1}\phi=-D^n(t\phi)=-[tD^n\phi+nD^{n-1}\phi]$$

from which, by means of (1),

$$H_{n+1}=tH_n-nH_{n-1}. \tag{3}$$

Also, from (1), $\phi DH_n + H_n D\phi = (-1)^n D^{n+1}\phi$,

from which $$H_{n+1} = tH_n - DH_n. \tag{4}$$

Hence, from (3), $$DH_n = nH_{n-1}. \tag{5}$$

By means of (3) or (4), and including the earlier results, we have

$$\left.\begin{aligned} H_0 = 1, \quad H_1 = t, \quad H_2 = t^2 - 1, \\ H_3 = t^3 - 3t, \quad H_4 = t^4 - 6t^2 + 3, \\ H_5 = t^5 - 10t^3 + 15t, \quad H_6 = t^6 - 15t^4 + 45t^2 - 15. \end{aligned}\right\} \tag{6}$$

Clearly, any polynomial is of the form

$$H_n = t^n + \alpha t^{n-2} + \ldots, \tag{7}$$

and hence $$D^n H_n = n!. \tag{8}$$

Multiply the expression for $\psi(t)$ in 7·06 (8) by H_r and integrate; then

$$\int_{-\infty}^{\infty} H_r \psi(t)\, dt = \sum_0 \int_{-\infty}^{\infty} a_n H_r D^n \phi\, dt. \tag{9}$$

Consider the integral

$$I_{r,n} \equiv \int_{-\infty}^{\infty} H_r D^n \phi\, dt = (-1)^n \int_{-\infty}^{\infty} H_r H_n \phi\, dt. \tag{10}$$

Then, from the first form of the integral,

$$I_{r,n} = [H_r D^{n-1}\phi]_{-\infty}^{\infty} - \int_{-\infty}^{\infty} DH_r D^{n-1}\phi\, dt.$$

The integrated part is equal to $(-1)^{n-1}[H_r H_{n-1}\phi]_{-\infty}^{\infty}$ which vanishes, since $H_r H_{n-1} = t^{r+n-1} + \ldots$ and

$$\lim_{|t| \to \infty} (t^p e^{-\frac{1}{2}t^2}) = 0.$$

Hence $$I_{r,n} = -\int_{-\infty}^{\infty} DH_r D^{n-1}\phi\, dt.$$

By repeated applications of this procedure we obtain

$$I_{r,n} = (-1)^n \int_{-\infty}^{\infty} D^n H_r \phi\, dt. \tag{11}$$

If n and r are unequal, the second form of the integral in (10) shows that it is immaterial as to which is the greater; let $n > r$; then, from (7), $D^n H_r = 0$ and hence, by (10) and (11),

$$\int_{-\infty}^{\infty} H_r D^n \phi\, dt = 0 \quad (r \neq n).$$

Accordingly, (9) reduces to

$$\int_{-\infty}^{\infty} H_r \psi(t)\,dt = a_r \int_{-\infty}^{\infty} H_r D^r \phi\,dt \equiv a_r I_{r,r}$$

$$= (-1)^r a_r \int_{-\infty}^{\infty} D^r H_r \phi\,dt$$

by (11). But, by (8), $D^r H_r = r!$; hence a_r is given by

$$\int_{-\infty}^{\infty} H_r \psi(t)\,dt = (-1)^r a_r r!. \tag{12}$$

7·08. Expression of the constants a_n and A_n in terms of the principal moments

By 7·06 (10),

$$\mu_r/\sigma^r = \int_{-\infty}^{\infty} t^r \psi(t)\,dt. \tag{1}$$

Now, by 7·07 (6) we can express t^r in terms of $H_r, H_{r-2}, \ldots$. Thus, with $H_0 = 1$, we have

$$\left.\begin{aligned} t &= H_1, \\ t^2 &= H_2 + H_0, \\ t^3 &= H_3 + 3H_1, \\ t^4 &= H_4 + 6H_2 + 3H_0, \\ t^5 &= H_5 + 10H_3 + 15H_1, \\ t^6 &= H_6 + 15H_4 + 45H_2 + 15H_0, \end{aligned}\right\} \tag{2}$$

or, in general,

$$t^r = H_r + B_{r,2} H_{r-2} + B_{r,4} H_{r-4} + B_{r,6} H_{r-6} + \cdots,$$

where, for values of $r = 1, 2, \ldots, 6$, the B's are given by the numerical coefficients in (2). Hence (1) becomes

$$\mu_r/\sigma^r = \int_{-\infty}^{\infty} (H_r + B_{r,2} H_{r-2} + \ldots)\,\psi\,dt$$

$$= (-1)^r [a_r r! + B_{r,2} a_{r-2} (r-2)! + B_{r,4} a_{r-4} (r-4)! + B_{r,6} a_{r-6} (r-6)! + \ldots]$$

by means of 7·07 (12).

We obtain the following results from the previous equation.

(i) $$\mu_0 \equiv \int_{-\infty}^{\infty} \psi(t)\,dt = a_0.$$

Since the value of the integral is 1, by 7·06 (9), then

$$a_0=\mu_0=1.$$

(ii) $$\mu_1/\sigma=-a_1.$$

But, by 7·05 (5), $\mu_1=0$; hence $a_1=0$.

(iii) $$\mu_2/\sigma^2=a_2.2!+a_0.$$

But $\mu_2\equiv\sigma^2$ and $a_0=1$; hence $a_2=0$.

(iv) $$\mu_3/\sigma^3=-[a_3 3!+3a_1]$$
$$=-a_3 3!.$$

(v) $$\mu_4/\sigma^4=a_4 4!+6a_2 2!+3a_0$$
$$=a_4 4!+3.$$

(vi) $$\mu_5/\sigma^5=-[a_5 5!+10a_3 3!+15a_1]$$
$$=-a_5 5!+10\mu_3/\sigma^3.$$

(vii) $$\mu_6/\sigma^6=[a_6 6!+15a_4 4!+45a_2 2!+15a_0]$$
$$=a_6 6!+15\mu_4/\sigma^4-30.$$

These give a_r in terms of the principal theoretical moments. Now, by 7·06 (7),

$$A_n=\sigma^n a_n n!,$$

and we obtain from (i)–(vii) the following expressions for the A's:

$$\left.\begin{aligned} A_0=1,\quad A_1=A_2=0,\quad A_3=-\mu_3,\\ A_4=\mu_4-3\sigma^4,\quad A_5=-(\mu_5-10\mu_3\sigma^2),\\ A_6=\mu_6-15\mu_4\sigma^2+30\sigma^6.\end{aligned}\right\} \tag{3}$$

The Gram-Charlier series, defined by 7·05 (2), becomes

$$F(x)=f(x)+\sum_3 \frac{A_n}{n!}\frac{d^nf(x)}{dx^n}, \tag{4}$$

where A_3, A_4, ... are given by (3).

We can express $F(x)$ in terms of Hermite's polynomials as follows. Now, by 7·06 (5) and 7·07 (1),

$$\frac{d^nf(x)}{dx^n}=\frac{1}{\sigma^{n+1}}D^n\phi=\frac{(-1)^n}{\sigma^{n+1}}H_n(t)\,\phi(t).$$

Also, $f(x)=\dfrac{1}{\sigma}\phi(t)$; hence

$$F(x)=\left[1+\sum_3(-1)^n\frac{A_n}{\sigma^n n!}H_n(t)\right]f(x), \tag{5}$$

where $t=(x-\bar{x})/\sigma$.

7·09. The characteristics of the Gram-Charlier series

The characteristics of the distribution given by 7·08 (4) or (5) are $\bar{x}$, $\mu_2 \equiv \sigma^2$ and A_3, A_4, ..., the latter bring functions of σ and the higher moments μ_3, μ_4, ...; in the evaluation of A_3, A_4, ... these higher moments are to be identified with the calculated moments m_3, m_4, ...

We can replace A_3 and A_4 as characteristics by the skewness, $\beta_1^{\frac{1}{2}}$, and the kurtosis, β_2, introduced in § 1·09 and defined by

$$\beta_1^{\frac{1}{2}} = \frac{\mu_3}{\sigma^3} \quad \text{and} \quad \beta_2 = \frac{\mu_4}{\mu_2^2} = \frac{\mu_4}{\sigma^4}; \tag{1}$$

hence, by 7·08 (3),

$$\frac{A_3}{\sigma^3} = -\beta_1^{\frac{1}{2}} \quad \text{and} \quad \frac{A_4}{\sigma^4} = \beta_2 - 3.$$

7·10. Theoretical frequency distributions derived from an assumed differential equation

The differential equation, first introduced in the present connexion by K. Pearson† and studied by him in all its many ramifications, is

$$\frac{1}{y}\frac{dy}{dx} = \frac{a-x}{b_0 + b_1 x + bx^2}, \tag{1}$$

in which a, b_0, b_1 and b are four arbitrary constants. The general solution of (1) is denoted by $y = F(x)$.

Many of the frequency distributions encountered in practice resemble certain theoretical distributions which are derived from (1) by assigning various values to the four constants; for example, the mode M is given in general by $dy/dx = 0$, so that for many distributions $M = a$, the value of a in (1) being now identified with the mode of the observed distribution.

It will be assumed that, for a given statistical distribution, the principal characteristics have been found by the usual procedure described in Chapter 1; these characteristics are the mean, $\bar{x}$, and the *calculated* principal moments, m_r.

If now $y \equiv F(x)$ is the theoretical relative distribution corresponding to the observed distribution, the theoretical principal moments being μ_r, the constants in (1) are to be found by identifying μ_r with m_r so as to provide a sufficient number of equations from which the constants can be evaluated.

Let x_1 and x_2 define the range of the variable. It will be assumed, meanwhile, that $y = 0$ at the extremities of the range; exceptional cases will be dealt with as they arise. Since $F(x)$ refers to a relative

† *Phil. Trans.* A, **186** (1892), 343; A, **216** (1916), 429.

distribution, we have at once the condition that the area under the curve is unity, that is,

$$\mu_0 \equiv \int_{x_1}^{x_2} y\,dx = \int_{x_1}^{x_2} F(x)\,dx = 1. \tag{2}$$

We shall frequently refer to this condition simply by saying that the area under the curve is finite.

We assume at the outset that the origin is chosen so that the y-axis is the centroid vertical, that is, the axis through the mean abscissa of the given distribution; thus, the calculated principal moment, m_1, is zero and hence, on equating m_1 and μ_1,

$$\mu_1 \equiv \int_{x_1}^{x_2} xy\,dx = 0, \tag{3}$$

in which $y = F(x)$. Similarly,

$$\mu_r = \int_{x_1}^{x_2} x^r y\,dx. \tag{4}$$

It is important to remember that throughout the range y is positive, being zero at the ends of the range except in the special cases alluded to.

If x_2 denotes a positive number and x_1 a positive or negative number such that $|x_1| < x_2$, the relevant possible ranges of the variable are:

(i) $x_1 \leqslant x \leqslant x_2$; (ii) $x_1 \leqslant x \leqslant \infty$;

(iii) $-\infty \leqslant x \leqslant x_2$; (iv) $-\infty \leqslant x \leqslant \infty$.

7·11. Standard form of the differential equation

The choice of origin made in the previous section leads to a simplification of the differential equation 7·10 (1) which is now written as

$$(a - x)\,y\,dx = (b_0 + b_1 x + bx^2)\,dy. \tag{1}$$

Hence
$$\int_{x_1}^{x_2} (a - x)\,y\,dx = [(b_0 + b_1 x + bx^2)\,y]_{x_1}^{x_2} - \int_{x_1}^{x_2} (b_1 + 2bx)\,y\,dx \tag{2}$$

or
$$(a + b_1)\int_{x_1}^{x_2} y\,dx - (1 - 2b)\int_{x_1}^{x_2} xy\,dx = P,$$

where P denotes the integrated part in (2). By 7·10 (2) the first integral is unity, and by 7·10 (3) the second integral is zero; hence

$$a + b_1 = P \equiv [(b_0 + b_1 x + bx^2)\,y]_{x_1}^{x_2}. \tag{3}$$

In the case when the limits are finite and $y = 0$ for $x = x_1$ and for $x = x_2$, then $P = 0$. In all other cases, as we shall see later, we can still take $P = 0$, provided certain restrictions are imposed on the values of one or more of the four disposable constants.

In all cases, then, we take $P=0$ and hence, from (3),

$$b_1=-a,$$

The differential equation can now be written as

$$\frac{1}{y}\frac{dy}{dx}=\frac{a-x}{b_0-ax+bx^2}\equiv-\frac{b_1+x}{b_0+b_1x+bx^2}, \tag{4}$$

either form of which we can regard as Pearson's standard form.

7·12. The theoretical principal moments

We have, from 7·10 (2) and (3),

$$\mu_0=1 \quad \text{and} \quad \mu_1=0. \tag{1}$$

Multiply 7·11 (1) throughout by x^r, where r has one of the values 1, 2 or 3, replacing b_1 by $-a$; then integrate between the limits x_1 and x_2 which are omitted, for simplicity, from the subsequent integrals. Thus

$$a\int x^r y\,dx-\int x^{r+1}y\,dx=\int(b_0x^r-ax^{r+1}+bx^{r+2})\,dy$$

$$=[x^r(b_0-ax+bx^2)\,y]_{x_1}^{x_2}$$

$$-\int\{b_0rx^{r-1}-a(r+1)\,x^r+b(r+2)\,x^{r+1}\}\,y\,dx,$$

whence, by 7·10 (4),

$$b_0r\mu_{r-1}-ar\mu_r+\{b(r+2)-1\}\,\mu_{r+1}=Q, \tag{2}$$

where Q, the integrated part, is given by

$$Q=[x^r(b_0-ax+bx^2)\,y]_{x_1}^{x_2}. \tag{3}$$

Now, the calculated moments, m_r, are finite and consequently, because of the identification of μ_r with m_r, the theoretical moments in (2) are finite, as they are in the case when the range is finite and $y=0$ at each end of the range; then $Q=0$. In all other cases, when, for example, the range is infinite or when y is infinite at one or both ends of a finite range, the theoretical moments are finite if $Q=0$; as we shall see later, the condition

$$Q=0$$

requires, in general, the restriction of the values of one or more of the constants in the differential equation.

For simplicity the expression for Q will be frequently written without the limits.

In (2) put $r=1, 2$ and 3 in succession; then, by means of (1), we have

$$b_0+(3b-1)\,\mu_2=0,$$
$$2a\mu_2-(4b-1)\,\mu_3=0,$$
$$3b_0\mu_2-3a\mu_3+(5b-1)\,\mu_4=0.$$

These are three linear equations in b_0, a and b from which each of the latter can be found in terms of μ_2, μ_3 and μ_4. If A is defined by

$$A\mu_2^3=10\mu_2\mu_4-12\mu_3^2-18\mu_2^3, \tag{4}$$

it is easily found that

$$\left.\begin{aligned} b_0A\mu_2^3&=\mu_2(4\mu_2\mu_4-3\mu_3^2),\\ aA\mu_2^3&=-\mu_3(\mu_4+3\mu_2^2),\\ bA\mu_2^3&=2\mu_2\mu_4-3\mu_3^2-6\mu_2^3.\end{aligned}\right\} \tag{5}$$

If now μ_2, μ_3 and μ_4 are equated to the known numerical values of m_2, m_3 and m_4 derived from the actual statistical distribution, then b_0, a and b can be evaluated.

In terms of the standard deviation, σ, and the characteristics β_1 and β_2, defined, for example, in 7·09 (1), we have

$$\mu_2\equiv m_2=\sigma^2, \quad \mu_3\equiv m_3=\beta_1^{\frac{1}{2}}\sigma^3, \quad \mu_4\equiv m_4=\beta_2\sigma^4.$$

Then, from (5), we have

$$\left.\begin{aligned} b_0&=\frac{\sigma^2}{A}(4\beta_2-3\beta_1),\\ b_1\equiv -a&=\frac{\sigma\beta_1^{\frac{1}{2}}}{A}(\beta_2+3),\\ b&=\frac{1}{A}(2\beta_2-3\beta_1-6),\end{aligned}\right\} \tag{6}$$

where A, defined by (4), is now given by

$$A=10\beta_2-12\beta_1-18. \tag{7}$$

The constants, b_0, a and b in the differential equation can now be evaluated in terms of the characteristics σ, β_1 and β_2 of the statistical distribution.

The differential equation becomes

$$-\frac{1}{y}\frac{dy}{dx}=\frac{\sigma\beta_1^{\frac{1}{2}}(\beta_2+3)+Ax}{\sigma^2(4\beta_2-3\beta_1)+\sigma\beta_1^{\frac{1}{2}}(\beta_2+3)\,x+(2\beta_2-3\beta_1-6)\,x^2}. \tag{8}$$

It is to be remembered that β_1 and β_2 are positive quantities.

7·13. Skewness and inflexion

The mode, M, is given in general by $M=a$ corresponding to a maximum turning point for which $dy/dx=0$. In terms of Pearson's definition (§ 1·09) the skewness is $(\bar{x}-M)/\sigma$. Now, since the y-axis is the centroid vertical, then $\bar{x}=0$; hence the skewness is $-a/\sigma$ or b_1/σ, which may be written, from 7·12 (5) and (6), in the alternative forms

$$\frac{\mu_3(\beta_2+3)}{A\sigma^3} \quad \text{or} \quad \frac{\beta_1^{\frac{1}{2}}}{A}(\beta_2+3).$$

The skewness is positive or negative according as A and μ_3 have the same or opposite signs.

For possible points of inflexion we must have $d^2y/dx^2=0$. Change the origin to the mode and write $x-a=\xi$; then the differential equation becomes

$$-\frac{1}{y}\frac{dy}{d\xi}=\frac{\xi}{d+c\xi+b\xi^2}$$

or

$$(d+c\xi+b\xi^2)\frac{dy}{d\xi}=-\xi y, \tag{1}$$

where $\quad d=b_0+a^2(b-1) \quad$ and $\quad c=a(2b-1)$.

Differentiate (1) with respect to ξ; then, putting $d^2y/d\xi^2=0$, we have

$$\{c+(2b+1)\xi\}\frac{dy}{d\xi}=-y. \tag{2}$$

Since $dy/d\xi=0$ only at the mode we have, from (1) and (2) by division,

$$\xi^2=\frac{d}{b+1}. \tag{3}$$

Thus there are two points of inflexion equidistant from the mode if d and $(b+1)$ have the same sign, or if

$$(b+1)\{b_0+a^2(b-1)\}>0. \tag{4}$$

7·14. The theoretical distributions

The differential equation is

$$\frac{1}{y}\frac{dy}{dx}=\frac{a-x}{b_0-ax+bx^2}, \tag{1}$$

the origin being such that the y-axis is the centroid vertical. In (1) the constants a $(\equiv -b_1)$, b_0 and b are expressed, if required, in terms of σ, β_1 and β_2 by means of 7·12 (5).

The distribution depends on the numerical values of a, b_0 and b, or, alternatively, of σ, β_1 and β_2, and, in particular, on the roots of the quadratic equation

$$b_0 - ax + bx^2 = 0, \qquad (2)$$

to which we refer as the *auxiliary quadratic*.

We now discuss the various forms of the theoretical curves.

7·15. Curves with mode at origin

Here $a = b_1 = 0$ and hence $\beta_1 = 0$. We write (1) in the alternative forms

$$-\frac{1}{y}\frac{dy}{dx} = \frac{x}{b_0 + bx^2} = \frac{Ax}{4\beta_2\sigma^2 + 2(\beta_2 - 3)x^2}, \qquad (1)$$

where

$$A = 10\beta_2 - 18. \qquad (2)$$

There are several cases according as b is zero or positive or negative.

(i) *$b = 0$ or $\beta_2 = 3$*

Then $A = 12$ and $b_0 \equiv 4\beta_2\sigma^2/A = \sigma^2$; hence

$$\frac{1}{y}\frac{dy}{dx} = -\frac{x}{\sigma^2},$$

and, consequently,

$$y = y_0 e^{-x^2/2\sigma^2},$$

where y_0 is the value of y when $x = 0$.

This is the normal distribution in which $y_0 = \dfrac{1}{\sigma\sqrt{(2\pi)}}$, since the area under the curve is unity.

Since

$$\lim_{|x| \to \infty} \{x^r e^{-x^2/2\sigma^2}\} = 0 \quad (r = 0, 1, \ldots, 4),$$

the integrated part, P, in 7·11 (2) vanishes and the integrated part, Q, in 7·12 (2) also vanishes.

The condition for points of inflexion, given by 7·13 (4), is satisfied and these are given by

$$\xi^2 \equiv x^2 = b_0 = \sigma^2,$$

as in § 1·15 (i), since $h^2 = 1/2\sigma^2$.

(ii) *b positive and $\beta_2 > 3$*

Then, by (2), $A > 0$ and hence $b_0 > 0$. From (1) we have at once

$$y = y_0\left(1 + \frac{b}{b_0}x^2\right)^{-1/2b}. \qquad (3)$$

The curve is symmetrical about the y-axis; also, since b_0 and b are positive, the range is $-\infty \leqslant x \leqslant \infty$ and the x-axis is an asymptote.

In (3) put $b_0 = b\alpha^2$ and $x = \alpha \tan\theta$. Then, since the area under the curve is unity, we have

$$2\alpha y_0 \int_0^{\frac{1}{2}\pi} \cos^{(1-2b)/b} \theta \, d\theta = 1.$$

Now, the Beta function, $B(m, n)$, is defined in general by 1·14 (15), namely,

$$B(m,n) = 2\int_0^{\frac{1}{2}\pi} \sin^{2m-1}\theta \cos^{2n-1}\theta \, d\theta, \tag{4}$$

with the conditions $m > 0$ and $n > 0$. Hence

$$\alpha y_0 B\left(\frac{1}{2}, \frac{1-b}{2b}\right) = 1, \tag{5}$$

in which, necessarily, $0 < b < 1$. This condition is automatically satisfied, for, with

$$\beta_1 = 0, \quad b = (2\beta_2 - 6)/A \quad \text{and} \quad A = 10\beta_2 - 18,$$

we obtain $\beta_2 > \frac{3}{2}$, which is in accordance with the initial assumption that $\beta_2 > 3$.

We can express (5) in terms of Gamma functions by means of 1·14 (16), namely,

$$B(m, n) = \frac{\Gamma(m) . \Gamma(n)}{\Gamma(m+n)}. \tag{6}$$

Hence, since $\Gamma(\frac{1}{2}) = \sqrt{\pi}$,

$$\alpha y_0 \sqrt{\pi}\, \Gamma\left(\frac{1-b}{2b}\right) \Big/ \Gamma\left(\frac{1}{2b}\right) = 1. \tag{7}$$

The value of y_0 is determined from (7) by means of tables of the Gamma function.

The range of values of b, however, is restricted more severely than that already derived for, to satisfy $Q = 0$ in 7·12 (2), we must have

$$\lim_{|x| \to \infty} \left\{ y_0 b_0 \left(1 + \frac{b}{b_0} x^2\right)^{1-1/(2b)} x^r \right\} = 0$$

for the values $r = 0$ (corresponding to P) and $r = 1$, 2 and 3. The condition is equivalent to

$$\lim_{|x| \to \infty} \{x^{2-(1/b)+r}\} = 0;$$

hence $b < 1/(2+r)$. For $r = 3$, the largest value under consideration, we have $b < \frac{1}{5}$.

It is easily seen that the condition for points of inflexion is satisfied; these are given by

$$x^2 = b_0/(b+1) = \frac{\beta_2 \sigma^2}{3(\beta_2 - 2)}.$$

The curve is shown in Fig. 14.

(iii) *b negative and* $0<\beta_2<3$

From 7·12 (6), A is positive; hence, from (2), $1\cdot 8<\beta_2<3$; also, from 7·12 (6), $b_0>0$.

In (1) write $\gamma^2=2(3-\beta_2)$ and $\gamma^2 q^2=4\beta_2\sigma^2$; then

$$-\frac{1}{y}\frac{dy}{dx}=\frac{Ax}{\gamma^2(q^2-x^2)}$$

and
$$y=y_0\left(1-\frac{x^2}{q^2}\right)^{A/(2\gamma^2)}.$$

Since $A>0$, the range is $-q\leqslant x\leqslant q$, and $y=0$ at each end of the range.

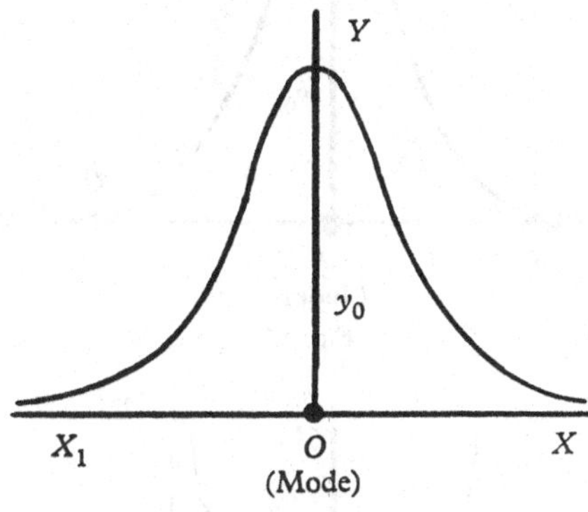

Fig. 14

The area under the curve is obviously finite.

Since the range is finite, the condition $Q=0$ is satisfied.

The condition for points of inflexion reduces to $(1+b)\,b_0>0$, or, since $b_0>0$, to $1+b>0$, from which $2<\beta_2<3$.

It is easily seen that y_0 is given by

$$y_0 qB\left(1+\frac{A}{2\gamma^2},\frac{1}{2}\right)=1.$$

Now,
$$\frac{dy}{dx}=-\frac{Axy_0}{q^2\gamma^2}\left(1-\frac{x^2}{q^2}\right)^{A/(2\gamma^2)-1}.$$

Hence the curve will touch the x-axis at each end of the range if $A>2\gamma^2$, that is, if $3>\beta_2>2\frac{1}{7}$. If $\beta_2=2\frac{1}{7}$, the tangent at each end of the range will make an acute angle $\tan^{-1}(Ay_0/q\gamma^2)$ or $\tan^{-1}(2qy_0/5\sigma^2)$ with the x-axis. If $1\cdot 8<\beta_2<2\frac{1}{7}$, the tangent at each end of the range will be perpendicular to the x-axis.

The curve for $2\frac{1}{7}<\beta_2<3$ is shown in Fig. 15.

(iv) *b positive and* $0<\beta_2<3$

In (i)–(iii) the initial assumption has been that the mode is at the origin, that is to say, y is a *maximum* when $x=0$. Although the present case does not conform to this assumption, it is convenient to consider it at this stage for, as we shall see, y is a *minimum* when $x=0$.

The formulae 7·12 (6) show that A is negative, so that $0<\beta_2<1{\cdot}8$, and that b_0 is negative.

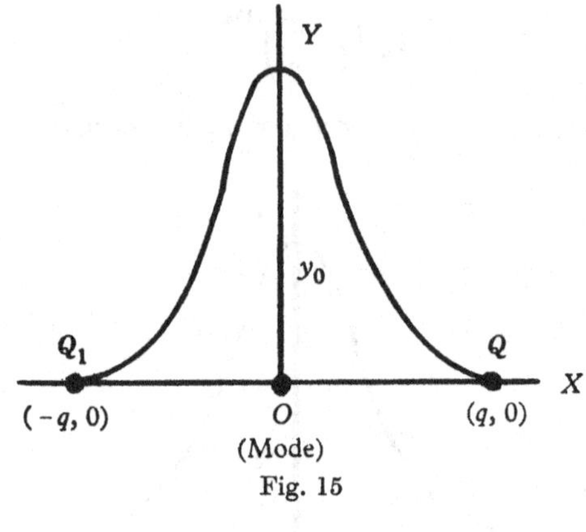

Fig. 15

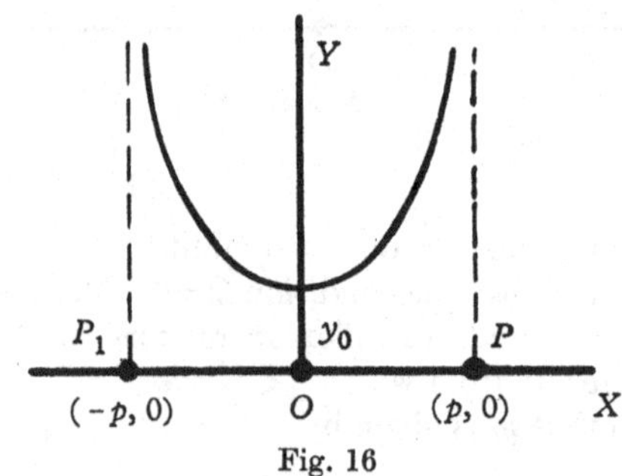

Fig. 16

From (3), on setting $p^2=-b_0/b$, we have

$$\frac{1}{y}\frac{dy}{dx}=\frac{x}{b(p^2-x^2)},$$

from which

$$y=y_0\left(1-\frac{x^2}{p^2}\right)^{-1/2b}.$$

The range is $-p\leqslant x\leqslant p$, and, at the ends of the range, $y=\infty$. The curve is shown in Fig. 16. It is obvious that y is a *minimum* when $x=0$; also, $x=\pm p$ are asymptotes.

In this case $\quad Q \equiv [x^r(b_0+bx^2)\,y]$

$$= -bp^2y_0\left[x^r\left(1-\frac{x^2}{p^2}\right)^{1-1/(2b)}\right];$$

hence $Q=0$ when $|x|=p$, if $1-1/(2b)>0$, that is, if $b>\frac{1}{2}$. This is the necessary condition for equating theoretical with calculated moments.

y_0 is given by

$$py_0B\left(\frac{1}{2}, 1-\frac{1}{2b}\right)=1.$$

7·16. Curves when $b=0$

From 7·12 (6) and (7), the given condition leads to the following:

$$2\beta_2=3\beta_1+6, \quad A=3(\beta_1+4), \quad b_0=\sigma^2, \quad b_1=\tfrac{1}{2}\sigma\beta_1^{\frac{1}{2}}. \tag{1}$$

Thus b_0 and b_1 are positive. Then

$$-\frac{1}{y}\frac{dy}{dx}=\frac{b_1+x}{b_1(x+\gamma)},$$

where $b_0=b_1\gamma$; γ is positive. Hence

$$y=y_0\left(1+\frac{x}{\gamma}\right)^f e^{-x/b_1}, \tag{2}$$

where

$$f=(\gamma-b_1)/b_1=(4-\beta_1)/\beta_1. \tag{3}$$

There are two cases according as $f>0$ or as $f<0$, that is, according as $\beta_1<4$ or as $\beta_1>4$. We ignore the trivial case when $f=0$.

(i) $f>0$ *or* $\beta_1<4$

The range is $-\gamma \leqslant x \leqslant \infty$. Since $f>0$, then $\gamma>b_1$; accordingly, the mode, $x=a=-b_1$, is on the origin side of the point $(-\gamma, 0)$.

Also,

$$Q\equiv[x^r(b_0+b_1x)\,y]$$

$$=b_1\gamma y_0\left[\left(1+\frac{x}{\gamma}\right)^{f+1}x^r e^{-x/b_1}\right]; \tag{4}$$

hence $Q=0$ at each end of the range.

By setting $x+\gamma=b_1t$ in (2), it is readily found by considering the area under the curve that

$$y_0 b_1\left(\frac{b_1}{\gamma}\right)^f e^{\gamma/b_1}\,\Gamma(1+f)=1. \tag{5}$$

From (2), $$\frac{dy}{dx}=\left(1+\frac{x}{\gamma}\right)^{f-1}R(x),$$

where $R(x)$ is finite when $x=-\gamma$. The x-axis is a tangent to the curve at $C(-\gamma, 0)$ if $f>1$, that is, if $0<\beta_1<2$. The curve is shown in Fig. 17.

If $f=1$, that is, if $\beta_1=2$, the tangent at C makes a finite acute angle with the x-axis. If $0<f<1$, that is, if $2<\beta_1<4$, the tangent at C is perpendicular to the x-axis.

(ii) $f<0$ *or* $\beta_1>4$

The range is $-\gamma \leqslant x \leqslant \infty$. Then $y=\infty$ when $x=-\gamma$ and $y=0$ when $x=\infty$. Also, from (3), $f+1=4/\beta_1$ and hence $f+1>0$; accordingly, from (4), $Q=0$ at each end of the range.

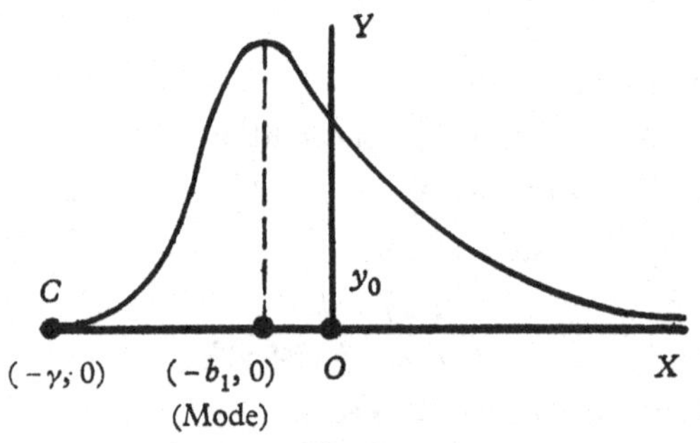

Fig. 17

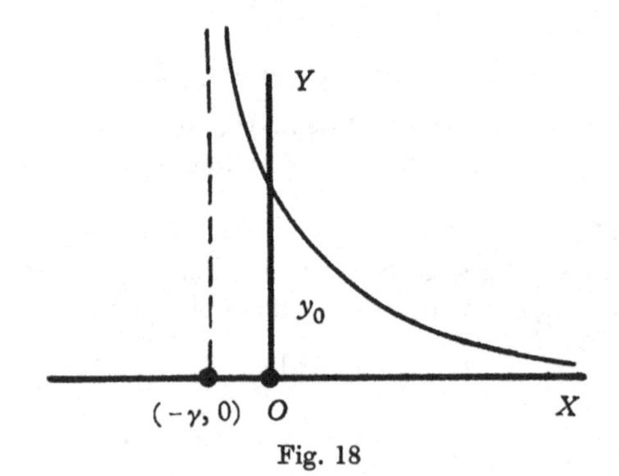

Fig. 18

The equation of the curve is given by (2) and y_0 is given by (5). The x-axis and the line $x=-\gamma$ are asymptotes. The curve is shown in Fig. 18. The mode is non-existent.

7·17. Curve when the roots of the auxiliary quadratic are imaginary

The differential equation can be written

$$\frac{1}{y}\frac{dy}{dx}=\frac{a-x}{b(x^2+2cx+d)}=\frac{a+c-(x+c)}{b\{(x+c)^2+g^2\}},$$

where $$2bc = -a, \quad bd = b_0 \quad \text{and} \quad g^2 = d - c^2. \tag{1}$$
Hence

$$\log\left(\frac{y}{y_0}\right) = -\frac{1}{2b}\log\left\{\frac{(x+c)^2+g^2}{c^2+g^2}\right\} + \frac{a+c}{bg}\left\{\tan^{-1}\left(\frac{x+c}{g}\right) - \tan^{-1}\left(\frac{c}{g}\right)\right\},$$

from which

$$y = y_0\left\{\frac{(x+c)^2+g^2}{c^2+g^2}\right\}^{-1/(2b)} \exp\left\{\frac{a+c}{bg}\tan^{-1}\left(\frac{gx}{cx+d}\right)\right\}. \tag{2}$$

Since $$\tan^{-1}\left(\frac{gx}{cx+d}\right) \to \tan^{-1}\left(\frac{g}{c}\right) \quad \text{as} \quad |x| \to \infty,$$

the range is infinite at both ends provided that $b > 0$; then $y = 0$ at each end of the range.

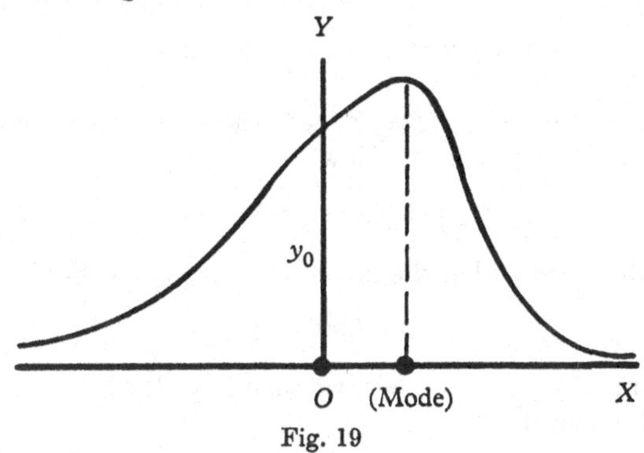

Fig. 19

From (1), $d > 0$; hence $b_0 > 0$.

The area under the curve is finite if $0 < b < 1$. Also, it is clear that $y \neq y_0$ at the mode $x = a$.

For finite moments, the restriction on the value of b is more severe than that just obtained. For large values of $|x|$,

$$Q \sim K[x^r\{(x+c)^2+g^2\}^{1-1/(2b)}],$$

and $Q = 0$ if $r + 2 < 1/b$ or, with $r = 3$, $b < \frac{1}{5}$.

The curve is shown in Fig. 19 when $a > 0$.

7·18. Curve when the roots of the auxiliary quadratic are equal

With the notation of §7·17, $g = 0$ and hence $d > 0$. Also

$$\frac{1}{y}\frac{dy}{dx} = \frac{a-x}{b(x+c)^2} = -\frac{1}{b(x+c)} + \frac{a+c}{b(x+c)^2},$$

from which $$y = y_0 c^{-1/b} e^{k/c} (x+c)^{-1/b} \exp\{-k/(x+c)\}, \tag{1}$$

where $k = (a+c)/b$.

Put $x+c = k/t^2$; then $$y = Ct^{2/b} e^{-t^2}, \tag{2}$$

where $$C = y_0 \left(\frac{c}{k}\right)^{1/b} e^{k/c}. \tag{3}$$

From (2), $y=0$ when $t=0$—that is, when $x=\infty$—provided that $b>0$.

From (1), $y \to 0$ when $x \to -c$, $(x+c>0)$, provided that $k>0$, that is, $a+c>0$.

The range is $$-c \leqslant x \leqslant \infty.$$

Also, since the y-axis is the centroid vertical, then $c>0$. Now, from 7·17 (1), $2bc = -a$; hence a is negative. Since $a+c>0$, then

$$2ab + 2bc > 0 \quad \text{or} \quad a(2b-1) > 0;$$

hence $0 < b < \frac{1}{2}$.

The values of b are more restricted when we apply the condition for finiteness of the moments. Now

$$Q = [bx^r(x+c)^2 y],$$

so that at the lower end of the range $(x=-c)$, $Q=0$. For large values of x,

$$Q \sim Kx^{2+r-1/b},$$

and $Q=0$ if $b < 1(2+r)$; for $r=3$, we then have $b<\frac{1}{5}$.

It is easily seen that

$$\frac{dy}{dx} = -\frac{C}{bk}(1-bt^2)\, t^{2+2/b} e^{-t^2}.$$

Hence the gradient is zero when $t=0$, $t=\infty$ and $t^2 = 1/b$, that is, when $x=\infty$, $x=-c$ and $x=a$, the last verifying that the mode is $x=a$.

Since

$$\int_{-c}^{\infty} y\, dx = 1,$$

we have from (2) $$2kC \int_0^{\infty} t^{(2/b)-3} e^{-t^2} dt = 1$$

or $$kC\Gamma\left(\frac{1}{b}-1\right) = 1.$$

This last result, together with (3), enables y_0 to be found in terms of k, b and c.

The curve is shown in Fig. 20.

Note. The above results can be deduced from the equations in § 7·17 by taking the limit when $g=0$ so that $d=c^2$. Thus, on reference to 7·17 (2),

$$\lim_{g\to 0}\left[\frac{a+c}{bg}\tan^{-1}\left\{\frac{gx}{c(x+c)}\right\}\right]=\frac{kx}{c(x+c)}=\frac{k}{c}-\frac{k}{x+c},$$

and (1) follows.

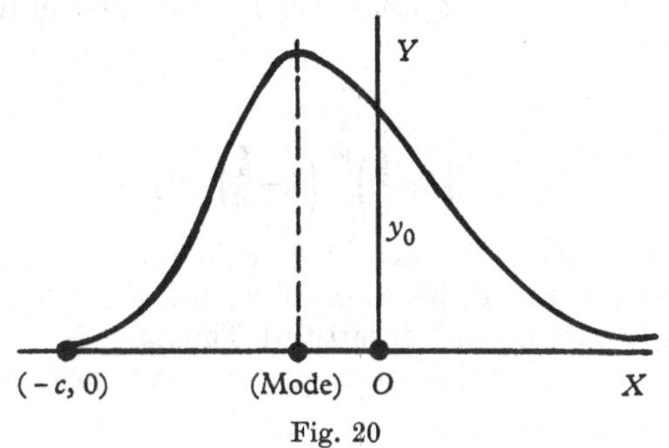

Fig. 20

7·19. Curves when the roots of the auxiliary quadratic are unequal

It is convenient to change the origin by writing $\xi=x-a$; then the differential equation is

$$\frac{1}{y}\frac{dy}{d\xi}=-\frac{\xi}{b\xi^2+B_1\xi+B_0}, \tag{1}$$

where, with the usual notation,

$$B_1=a(2b-1),\quad B_0=b_0+a^2(b-1). \tag{2}$$

The new origin is the mode if this exists; in several cases, as we shall see, the mode is non-existent.

Let α and β be the roots of the new auxiliary quadratic; then

$$b(\alpha+\beta)=-B_1,\quad b\alpha\beta=B_0. \tag{3}$$

It will be sufficient to consider the general case when bB_1 is negative so that $\alpha+\beta$ is positive. We take β to be positive; then α may be positive or negative, with $|\alpha|<\beta$.

From (1),

$$\frac{1}{y}\frac{dy}{d\xi}=\frac{p\alpha}{\xi-\alpha}-\frac{p\beta}{\xi-\beta},$$

where

$$p=\frac{1}{b(\beta-\alpha)};$$

then p has the sign of b.

The differential equation takes the two distinctive forms:

$$\frac{1}{y}\frac{dy}{d\xi}=-\frac{p\alpha}{\alpha-\xi}+\frac{p\beta}{\beta-\xi} \tag{4}$$

and

$$\frac{1}{y}\frac{dy}{d\xi}=\frac{p\alpha}{\xi-\alpha}+\frac{p\beta}{\beta-\xi}. \tag{5}$$

In (i)–(iii) following we consider the form (4) and in (iv)–(vi) the form (5).

(i) $0<\alpha<\beta,\ b>0$

From (4),

$$y=y_0\left(1-\frac{\xi}{\alpha}\right)^{p\alpha}\left(1-\frac{\xi}{\beta}\right)^{-p\beta}, \tag{6}$$

where $y=y_0$ when $\xi=0$, that is, at the mode if it exists.

Here, $p>0$ and $p\alpha>0$; hence $y=0$ when $\xi=\alpha$. Also, $y=0$ when $\xi=-\infty$ if $p(\beta-\alpha)>0$, which is satisfied. The range is $-\infty\leqslant\xi\leqslant\alpha$.

For finite area under the curve we must have $p(\beta-\alpha)>1$; thus, $0<b<1$.

In terms of ξ, $Q=[b(\xi-\alpha)(\xi-\beta)(\xi+a)^r y]$.

At the upper end of the range ($\xi=\alpha$), $Q=0$. At the lower end $Q=0$ if $2+r-p(\beta-\alpha)<0$; thus, with $r=3$, $0<b<\frac{1}{5}$.

From (2) and (3), $a=b(\alpha+\beta)/(1-2b)$; hence a is positive.

From (6),

$$\frac{dy}{d\xi}=-\frac{y_0\xi}{b\alpha\beta}\left(1-\frac{\xi}{\alpha}\right)^{p\alpha-1}\left(1-\frac{\xi}{\beta}\right)^{-p\beta-1}.$$

The x-axis will be a tangent at $A(\alpha,0)$ if $p\alpha>1$, that is, if

$$\beta/\alpha<1+1/b.$$

The gradient of the tangent to the curve at A will be finite and non-zero if $p\alpha=1$, that is, if $\beta/\alpha=1+1/b$. The tangent to the curve at A will be perpendicular to the x-axis if $p\alpha<1$, that is, if $0<\beta/\alpha<1+1/b$.

The curve is shown in Fig. 21 in which CV is the centroid vertical, O_1 is the origin for the (ξ,η) coordinates and $p\alpha>1$.

(ii) $\beta>0,\ \alpha<0,\ \beta+\alpha>0;\ b<0$

Here p is negative. Write $\alpha_1=-\alpha$ and $p_1=-p$, where α_1 and p_1 are positive; then $\beta>\alpha_1$.

From (4),

$$y=y_0\left(1+\frac{\xi}{\alpha_1}\right)^{p_1\alpha_1}\left(1-\frac{\xi}{\beta}\right)^{p_1\beta}. \tag{7}$$

The range is $-\alpha_1\leqslant\xi\leqslant\beta$. The moments are finite.

From (3), $B_1>0$ and $B_0>0$; hence, from (2), $a<0$ and $b_0>0$.

From (7),

$$\frac{dy}{d\xi}=\frac{y_0\xi}{b\alpha_1\beta}\left(1+\frac{\xi}{\alpha_1}\right)^{p_1\alpha_1-1}\left(1-\frac{\xi}{\beta}\right)^{p_1\beta-1}.$$

If $p_1\alpha_1 > 1$, the gradients at $\xi = -\alpha_1$ and $\xi = \beta$ are zero; if $p_1\alpha_1 = 1$, the gradient at $\xi = -\alpha_1$ is positive and finite, and the gradient at $\xi = \beta$ is zero; if $p_1\alpha_1 < 1$ the gradient at $\xi = -\alpha_1$ is infinite and the gradient at $\xi = \beta$ is zero, finite, or infinite according as $p_1\beta > 1$, $= 0$, or < 1.

The curve is shown in Fig. 22 when $p_1\alpha_1 > 1$.

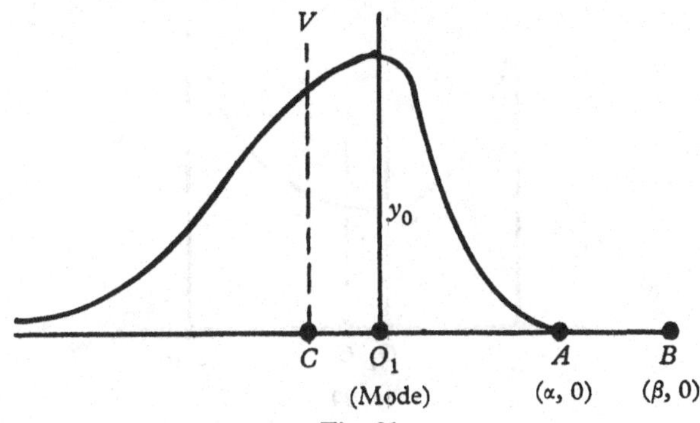

Fig. 21

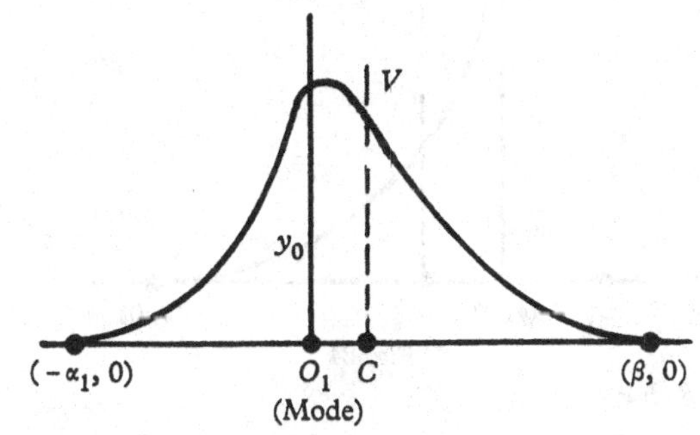

Fig. 22

(iii) $\beta > 0$, $\alpha < 0$; $b > 0$

From (4), with $\alpha_1 = -\alpha$,

$$y = y_0\left(1 + \frac{\xi}{\alpha_1}\right)^{-p\alpha_1}\left(1 - \frac{\xi}{\beta}\right)^{-p\beta}.$$

Here $p > 0$ and the range is $-\alpha_1 \leqslant \xi \leqslant \beta$. The ordinates at the ends of the range are asymptotes.

The differential equation shows that $dy/d\xi = 0$ when $\xi = 0$; the point $(0, y_0)$ is a *minimum* turning-point.

We consider the case when $\alpha_1 < \beta$; the procedure is similar when $\alpha_1 > \beta$. Then, from (3), $B_1 < 0$ and $B_0 < 0$; from (2), $a > 0$ if $0 < b < \frac{1}{2}$ and $a < 0$ if $b > \frac{1}{2}$.

The area under the curve is finite if $0<p\alpha_1<p\beta<1$. It is easily seen that the moments are finite for, in terms of ξ,

$$Q=-b\alpha_1\beta y_0\left[(\xi+a)^r\left(1+\frac{\xi}{\alpha_1}\right)^{1-p\alpha_1}\left(1-\frac{\xi}{\beta}\right)^{1-p\beta}\right],$$

so that $Q=0$ at each end of the range.

The curve, when $\alpha_1<\beta$ and $b>\frac{1}{2}$, is shown in Fig. 23.

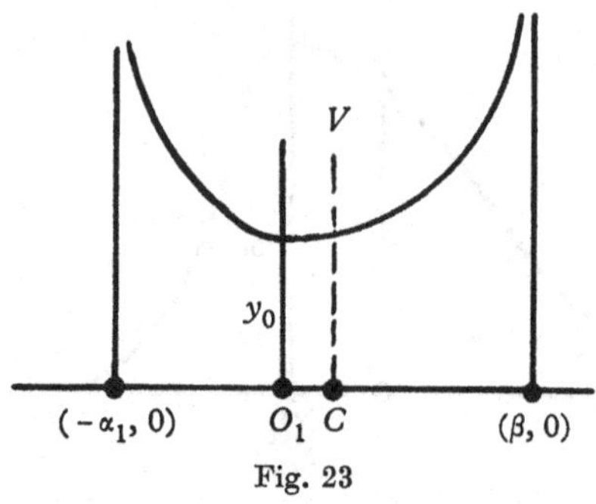

Fig. 23

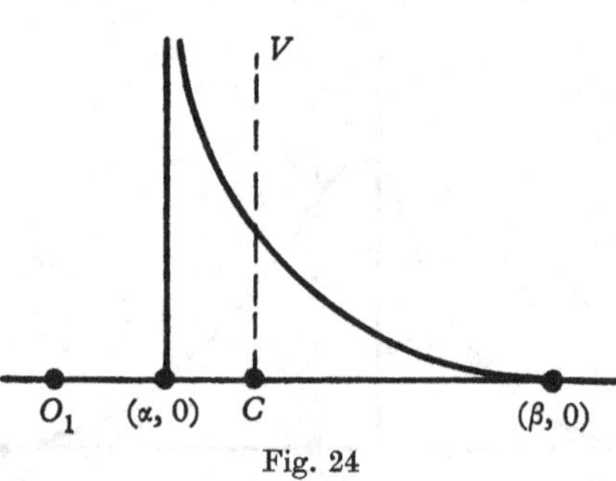

Fig. 24

(iv) $0<\alpha<\beta$, $b<0$

We now consider the form (5) of the differential equation. Here $p<0$. Then, with $p_1=-p$,

$$y=C(\xi-\alpha)^{-p_1\alpha}(\beta-\xi)^{p_1\beta},$$

where C is a constant.

The range is $\alpha\leqslant\xi\leqslant\beta$. The line $\xi=\alpha$ is an asymptote; also $y=0$ when $\xi=\beta$. Further, from (2) and (3), $a<0$.

The condition for finite area under the curve is $0<p_1\alpha<1$.

As in (iii) it is easily seen that $Q=0$ at each end of the range.

The gradient at $(\beta, 0)$ depends on the value of $\lim\limits_{\xi\to\beta}[(\beta-\xi)^{p_1\beta-1}]$. When $p_1\beta>1$, the gradient is zero. The other cases, $p_1\beta=1$ and $p_1\beta<1$, follow as before. The curve, when $p_1\beta>1$, is shown in Fig. 24. The mode is non-existent.

(v) $0<\alpha<\beta,\ b>0$

Here $p>0$ and, from (5),

$$y=C(\xi-\alpha)^{p\alpha}(\beta-\xi)^{-p\beta}.$$

The range is $\alpha\leqslant\xi\leqslant\beta$. The line $\xi=\beta$ is an asymptote. For finite area under the curve we must have $0<p\beta<1$. Also, from (3), $B_1<0$ and $B_0>0$. Since the centroid vertical lies between $\xi=\alpha$ and $\xi=\beta$, then $a<0$ and, if $\alpha_1=-\alpha$,

$$b(\alpha+\beta)=a_1(2b-1);$$

hence $b>\frac{1}{2}$. Also, $Q=0$ at each end of the range.

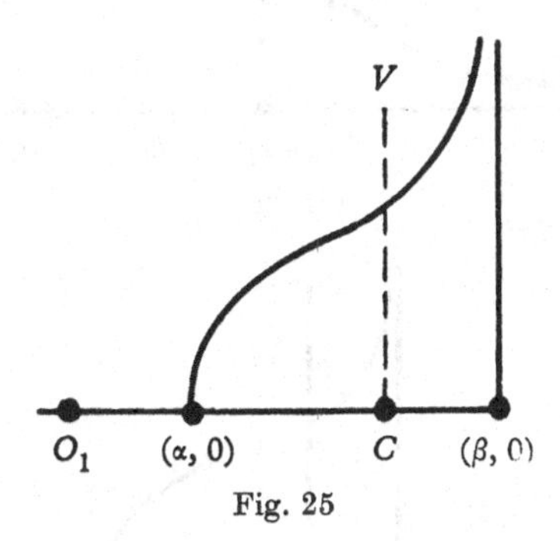

Fig. 25

The value of the gradient at $\xi=\alpha$ depends on $\lim\limits_{\xi\to\alpha}[(\xi-\alpha)^{p\alpha-1}]$.

Now, $p\alpha\equiv p\beta.\dfrac{\alpha}{\beta}$, and hence, since $0<p\beta<1$ and $\alpha<\beta$, then $p\alpha-1$ is negative; accordingly, the gradient at $(\alpha,0)$ is infinite.

The curve is shown in Fig. 25. The mode is non-existent.

(vi) $\alpha<0,\ \beta>0;\ b>0$

Here $p>0$ and, from (5) with $\alpha_1=-\alpha$,

$$y=C(\xi+\alpha_1)^{-p\alpha_1}(\beta-\xi)^{-p\beta}.$$

The range is $-\infty\leqslant\xi\leqslant-\alpha_1$. The requirements for finite area under the curve are $-p\alpha_1>-1$ and $-p(\alpha_1+\beta)<-1$ or $p\alpha_1<1$ and $p(\beta+\alpha_1)>1$. The x-axis and the line $\xi=-\alpha_1$ are asymptotes.

Taking the case $\beta>\alpha_1$ we find from (2) and (3) that $B_1<0$ and $B_0<0$. Also, since the centroid vertical must lie between $-\infty$ and $-\alpha_1$ then $a>0$ and consequently, from (2) and (3), $0<b<\frac{1}{2}$.

The curve is shown in Fig. 26.

(vii) $A=0$

We take the centroid vertical to be the y-axis.

From 7·12 (7), $5\beta_2=6\beta_1+9$. Then the differential equation, 7·12 (8), becomes

$$\frac{1}{y}\frac{dy}{dx}=\frac{q}{x^2-qx-3\sigma^2},$$

where $q=2\sigma\beta_1^{\frac{1}{2}}$.

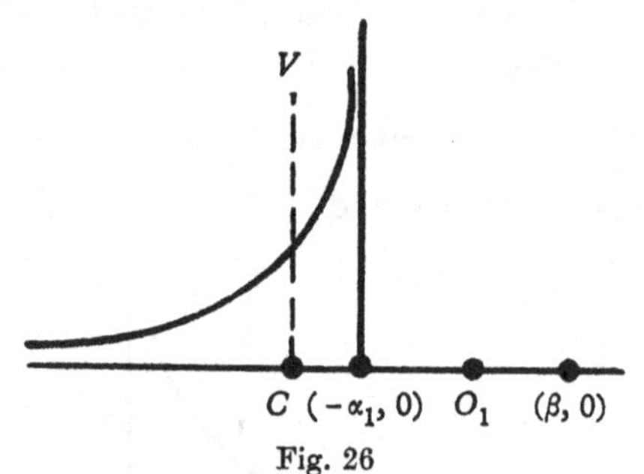

Fig. 26

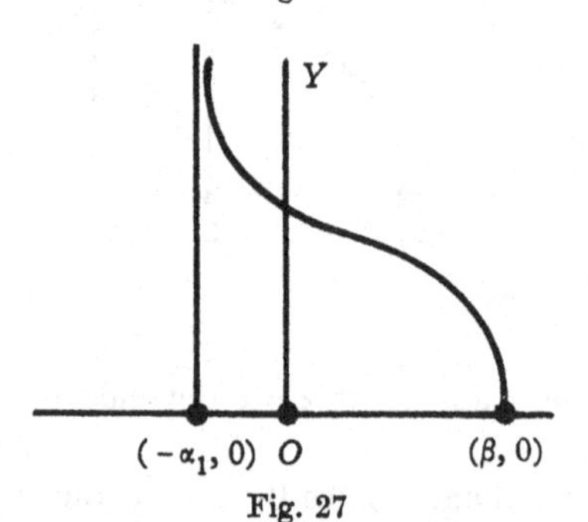

Fig. 27

We first consider the case when $\beta_1^{\frac{1}{2}}\equiv\mu_3/\sigma^3$ is positive; then $q>0$. The roots of the auxiliary equation are of opposite sign, say, β and $-\alpha_1$, where β and α_1 are positive. Also, since $\beta-\alpha_1=q$, then $\beta>\alpha_1$. We now obtain

$$y=y_0\left(1+\frac{x}{\alpha_1}\right)^{-p}\left(1-\frac{x}{\beta}\right)^{p},$$

where $p=q/(\beta+\alpha_1)$; thus, $p>0$. The range is $-\alpha_1\leqslant x\leqslant\beta$. Also, $y=0$ when $x=\beta$, and $y=\infty$ when $x=-\alpha_1$. The mode is non-existent, for it is easily seen that $dy/dx\neq 0$ for any point within the range.

For finite area under the curve we must have $0<p<1$ or $q\,(\equiv\beta-\alpha_1)<\beta+\alpha_1$, which is satisfied. Further, it is easily seen that $Q=0$.

The gradient at $(\beta,0)$ depends on the value of $\lim\limits_{x\to\beta}[(\beta-x)^{p-1}]$ and is consequently infinite.

The curve is shown in Fig. 27; there is no point of inflexion.

If q is negative, that is, if μ_3 is negative, we write $q=-q_1$ where $q_1>0$. Also, if $x=-x_1$, the equation (8) becomes

$$\frac{1}{y}\frac{dy}{dx_1}=\frac{q_1}{x^2-q_1x_1-3\sigma^2}.$$

The resultant curve is simply the reflexion of Fig. 27 in the y-axis.

7·20. Example of a Pearson curve

The data with which we shall be concerned are derived from experiments† made by Reed and McLeod on 'the absorption of underwater sound by substances in process of solution'. The time, t, in seconds, is the independent variable; for convenience, we write $X=4t$ in Table 17. The measured quantity is denoted by y, the values of which for a particular experiment were kindly supplied by Dr Reed. A smooth curve (Fig. 28) was drawn with reference to the points (X, y) and the values of y, read from the curve, are in the second column of Table 17. According to the experiment, $y=0$ when $t=X=0$; the curve indicates that y rapidly tends to zero beyond $X=15$.

The moments, $\mu_r(a)$, are first calculated with reference to the line $X=a$, where $a=6$; the deviations—denoted here by η to avoid confusion later—are given by

$$\eta=X-6, \tag{1}$$

and are shown in the third column of Table 17.

From the sums at the feet of the columns we have the following:

$$N=\Sigma y=33{\cdot}1,\quad [y\eta]=-11{\cdot}05,\quad [y\eta^2]=292{\cdot}25,$$
$$[y\eta^3]=190{\cdot}55,\quad [y\eta^4]=6814{\cdot}25.$$

Then

$$\bar{\eta}=\frac{[y\eta]}{N}=-0{\cdot}334,$$

$$\mu_2(a)=\frac{[y\eta^2]}{N}=8{\cdot}829,$$

$$\mu_3(a)=\frac{[y\eta^3]}{N}=5{\cdot}757,$$

$$\mu_4(a)=\frac{[y\eta^4]}{N}=205{\cdot}87.$$

The values of the principal moments are now calculated by means of the formulae 1·06 (12), (15) and (16); they are

$$\mu_2\equiv\sigma^2=8{\cdot}718,\quad \mu_3=14{\cdot}512,\quad \mu_4=219{\cdot}395.$$

† R. D. C. Reed and T. C. McLeod, *Nature, Lond.*, **175** (1955), p. 809.

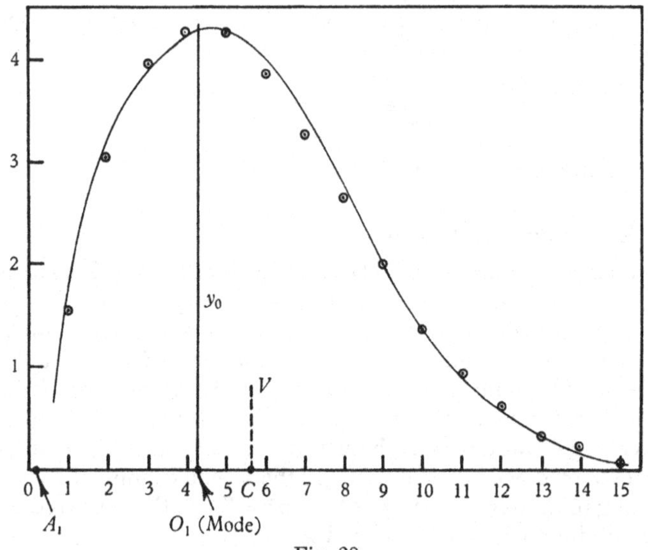

Fig. 28

Table 17. *Absorption of underwater sound*

X	y	η	yη +	yη −	yη²	yη³ +	yη³ −	yη⁴	y (calc.)
1	1·7	−5	—	8·5	42·5	—	212·5	1062·5	1·55
2	3·2	−4	—	12·8	51·2	—	204·8	819·2	3·02
3	3·85	−3	—	11·55	34·65	—	103·95	311·85	3·95
4	4·2	−2	—	8·4	16·8	—	33·6	67·2	4·27
5	4·25	−1	—	4·25	4·25	—	4·25	4·25	4·27
6	4·0	0	0	—	0	0	—	0	3·82
7	3·45	+1	3·45	—	3·45	3·45	—	3·45	3·27
8	2·8	+2	5·6	—	11·2	22·4	—	44·8	2·64
9	2·0	+3	6·0	—	18·0	54·0	—	162·0	2·01
10	1·4	+4	5·6	—	22·4	89·6	—	358·4	1·38
11	0·9	+5	4·5	—	22·5	112·5	—	562·5	0·98
12	0·6	+6	3·6	—	21·6	129·6	—	777·6	0·62
13	0·4	+7	2·8	—	19·6	137·2	—	960·4	0·35
14	0·25	+8	2·0	—	16·0	128·0	—	1024·0	0·25
15	0·1	+9	0·9	—	8·1	72·9	—	656·1	0·08
	33·1		**34·45**	**45·50**	**292·25**	**749·65**	**559·1**	**6814·25**	
			−11·05			**+190·55**			

From these $\beta_1^{\frac{1}{2}} \equiv \frac{\mu_3}{\sigma^3} = 0{\cdot}5636$ and $\beta_1 = 0{\cdot}3176$;

$$\beta_2 \equiv \frac{\mu_4}{\sigma^4} = 2{\cdot}8867.$$

Then, from 7·12 (7) and (6) we obtain the following:

$$A = 7{\cdot}056, \quad b_0 = 13{\cdot}10, \quad b_1 = 1{\cdot}389, \quad b = -0{\cdot}1674.$$

If x is the abscissa with reference to the centroid vertical (CV in Fig. 28), the auxiliary quadratic is

$$-0{\cdot}1674x^2 + 1{\cdot}389x + 13{\cdot}10 = 0$$

or

$$x^2 - 8{\cdot}299x - 78{\cdot}25 = 0, \tag{2}$$

of which the roots are

$$-5{\cdot}621 \quad \text{and} \quad 13{\cdot}920.$$

Since the mean, $\bar{\eta}$, is $-0{\cdot}334$, the position of CV is given, from (1), by $X = 5{\cdot}666$, that is, $OC = 5{\cdot}666$.

The root $x = -5{\cdot}621$ of (2) is associated with the point A_1 in Fig. 28, so that $A_1C = 5{\cdot}621$; thus A_1 is very close to the origin O, exact identity being hardly expected in the circumstances of an involved arithmetic computation.

In terms of the notation of the differential equation 7·11 (4) with CV as y-axis, the mode, O_1, is given by $x = a = -b_1$; thus, in Fig. 28, $O_1C = 1{\cdot}389$; also, $OO_1 = 4{\cdot}277$.

If ξ denotes the abscissa of a point on the curve with the mode, O_1, as origin then the roots of the auxiliary equation expressed in terms of ξ are

$$\alpha = -5{\cdot}621 + b_1 \quad \text{and} \quad 13{\cdot}920 + b_1$$

or

$$\alpha = -4{\cdot}232, \quad \beta = 15{\cdot}309.$$

The values of b, α and β show that the curve belongs to the category treated in general terms in § 7·19 (ii); thus, with the notation of this section,

$$\alpha_1 = 4{\cdot}232, \quad \beta - \alpha = 19{\cdot}541;$$

$$p_1 = \frac{1}{|b|(\beta - \alpha)} = \frac{1}{3{\cdot}271}; \quad p_1\alpha_1 = 1{\cdot}293 \quad \text{and} \quad p_1\beta = 4{\cdot}680.$$

From 7·19 (7), the equation of the curve is

$$y = y_0\left(1 + \frac{\xi}{\alpha_1}\right)^{p_1\alpha_1}\left(1 - \frac{\xi}{\beta}\right)^{p_1\beta}, \tag{3}$$

where y_0 is the maximum ordinate (corresponding to the mode).

In Fig. 28, $A_1O_1=4{\cdot}232$; if B_1 corresponds to the root β, then $O_1B_1=15{\cdot}309$ and $OB_1=19{\cdot}586$; B_1 lies outside the range of the figure.

Since $p_1\alpha_1>1$, the theoretical curve given by (3) touches the ξ-axis at A_1 and B_1.

Now, $OO_1=4{\cdot}277$; accordingly, the relation between the X and ξ abscissae is

$$X=\xi+4{\cdot}277.$$

Thus, when $X=1$, $\xi=-3{\cdot}277$; when $X=2$, $\xi=-2{\cdot}277$; and so on. The values of y, calculated from (3) with these values of ξ, that is, for $X=1, 2, \ldots$, are given in the final column of Table 17. The corresponding points are indicated by circles in Fig. 28.

It is seen from the figure, or by comparing the second and last columns of the table, that the theoretical curve (3) and the smoothed curve are in reasonably good agreement.

The formula (3) may then be regarded as summarizing the results of the experiment.

CHAPTER 8

CORRECTION OF STATISTICS

8·01. Possible rejection of an observation

In § 2·20 the normal law of errors was tested against the aggregate of residuals which, in the circumstances stated, may be regarded substantially as equivalent to the errors of observation; it was seen that there was a close correspondence between the theoretical distribution and the observed distribution of the residuals except in the case of a small number of large residuals, this number being somewhat greater than that predicted by the normal law, a feature not uncommon in a series of observations.

For purposes of explanation we shall suppose that all the measures or observations but one provide residuals in general accordance with the normal law and that the exceptional measure, M, leads to a residual which appears to be 'abnormally' large. The normal law, of course, allows for large errors, but the probability of their occurrence is, relatively, very small. There are two possibilities: first, the normal law may just be a very good approximation, with a slight failure to predict the frequency of very large errors; secondly, an exceptional measure, such as M, may be the result of some 'abnormal' combination of circumstances, such as would occur if the errors produced by the various agencies were all of the same sign. It may be that, after the residuals of all the measures have been obtained, the exceptional measure, M, can be repeated, as when measures of stellar images on a photographic plate are involved. On the other hand, M may represent a particular measure of the right ascension of a star observed with the meridian circle, in which case no scrutiny of the actual observation—which is now past recall—is possible, although the details of subsequent arithmetical computation can be checked.

When the number, n, of observations is small, say, $n=6$, the retention or suppression of an 'exceptional' measure will clearly lead to a considerable variation in the deduced arithmetical mean and in its standard or probable error. The computer, who is frequently not the observer, may then be greatly tempted to reject what he imagines to be a 'discordant' observation, forgetting, or perhaps being unaware, that frequency is not strictly related to probability when the number of measures is small. The practice of experienced observers is to *retain* an apparently discordant measure when there are no indications that the circumstances in which the observation was made were other than normal; in other words, personal prejudice should not be exercised, when once the observations have been made, in suggesting the rejection of a particular measure.

On the other hand, it has been claimed that a discordant measure is the victim of abnormal circumstances, and, for this reason, should be rejected provided that the degree of discordance is subject to some criterion. Such a criterion, elaborate in its application, was first proposed by Peirce†; a simpler criterion for the rejection of *one* doubtful observation was later given by Chauvenet (op. cit. p. 564), and is described in the next section.

8·02. Chauvenet's criterion

If h is the modulus of precision of n measures, the probability of the occurrence of an error x is $\operatorname{erf}(hx)$; hence the probability that an error will exceed the value x is $\{1-\operatorname{erf}(hx)\}$, so that in general, provided frequency is identified with probability, the number of measures with errors exceeding x is $E \equiv n\{1-\operatorname{erf}(hx)\}$. If $E < \frac{1}{2}$, then a measure with an error α $(\alpha > x)$ is to be rejected, on the grounds that the probability against its occurrence is greater than the probability in its favour. Thus the limit for rejection of *one* doubtful observation is given by

$$n\{1-\operatorname{erf}(hx)\} = \tfrac{1}{2},$$

from which

$$\operatorname{erf}(hx) = \frac{2n-1}{2n}. \tag{1}$$

With h and n known the tables of the erf function enable us to deduce the appropriate value of the limiting error x.

To illustrate the procedure we consider the example given by Chauvenet (op. cit. p. 562) relating to fifteen measures of the vertical semi-diameter of the planet Venus. The standard error, μ, computed from the residuals in the usual way, is found to be $0''\cdot572$; the residuals are $-1''\cdot40$ and fourteen others ranging between $-0''\cdot44$ and $+1''\cdot01$.

From (1), $\operatorname{erf}(hx) = 0\cdot96667$; from the tables of the erf function, $hx = 1\cdot5047$. Also, since $h^2 = 1/(2\mu^2)$, it is readily found that $x = 1''\cdot22$. There is only one residual—with the numerical value $1''\cdot40$—which exceeds this limiting value of x and, according to Chauvenet's criterion, the corresponding measure is to be rejected.

After this rejection has been made from the original series of measures and new residuals calculated, the criterion can be applied a second time and, possibly, another measure may fall under the ban of rejection.

† B. Peirce, *Astron. J.* **2** (1853), 161; see also, W. Chauvenet, *Spherical and Practical Astronomy* (Philadelphia, 1891), vol. II, p. 558.

Whatever the theoretical justification of Chauvenet's criterion may be—and it would appear to be none too strong in the case of a small number of observations, for these would not give full play to the arguments based on probability—it has the merit of being impersonal and not subservient to the bias of the individual. It should be regarded at best as an empirical rule, to be applied sparingly and with discretion.

One important point relates to the finally adopted degree of precision of the measures. In the example fifteen measures have been made and the precision of these is given by $\mu = 0''{\cdot}572$. This estimate of the precision should remain and be independent of the rejection of a 'discordant' measure, although the rejection leads to what is hoped to be a more acceptable value of the unknown quantity.

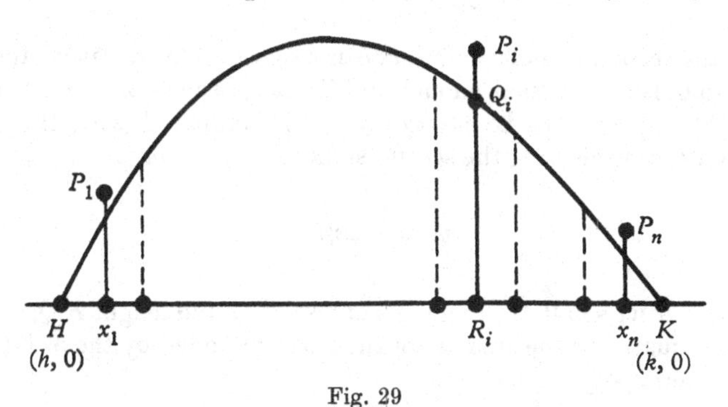

Fig. 29

8·03. Sheppard's corrections to moments

We suppose that for a given statistical distribution the range of the variable x is divided into n class intervals, each of width c, with their centres at the points whose abscissae are x_i $(i = 1, 2, \ldots, n)$, as in Fig. 29. We assume that the distribution extends between H and K whose abscissae are $h \equiv x_1 - \frac{1}{2}c$ and $k \equiv x_n + \frac{1}{2}c$. In the typical class interval the frequency is $R_i P_i$, corresponding to the vertex P_i, with abscissae x_i, of the frequency polygon.

We take the origin to be at the mean of the given distribution and we shall be concerned with *principal* moments.

We denote the *calculated* principal moments of order r by m_r, so that

$$m_r = \frac{1}{N} \sum_{i=1}^{n} R_i P_i . x_i^r, \tag{1}$$

where N is the total frequency. It is assumed that m_r is evaluated according to the procedures of Chapter 1; in particular, $m_0 = 1$ and $m_1 = 0$.

Let $y=f(x)$ be the equation of the curve in Fig. 29 representing the *true* distribution of the variable x such that $y=0$ when $x=h$ and $x=k$; also let $F(x)$ denote the true relative distribution function so that $f(x)=NF(x)$. Then the true principal moment of order r, denoted by μ_r, is given by

$$\mu_r=\int_h^k x^r F(x)\,dx. \tag{2}$$

The object of Sheppard's corrections is to deduce the values of the μ's from the calculated and known values of the m's.

Let the ordinate of P_i meet the curve at Q_i, so that $R_i Q_i=y_i\equiv f(x_i)$; then, from (1),

$$m_r=\frac{1}{N}\Sigma x_i^r y_i+\frac{1}{N}\Sigma x_i^r\,.\,Q_i P_i.$$

In the second summation, $Q_i P_i$ can be positive or negative and, if the subdivisions are small enough and N is large, the second summation may be expected to be negligibly small compared with the first summation; neglecting the second summation, we have

$$m_r=\frac{1}{N}\Sigma x_i^r y_i. \tag{3}$$

Now, with c small, the rectangle of breadth c and height R_iQ_i $(\equiv y_i)$ can be equated to the area under the curve bounded by the ordinates $x_i-\frac{1}{2}c$ and $x_i+\frac{1}{2}c$. Thus

$$cy_i=\int_{x_i-\frac{1}{2}c}^{x_i+\frac{1}{2}c} f(x)\,dx,$$

or, on writing $x\equiv x_i+\xi$ and $f(x)=NF(x)$,

$$cy_i=N\int_{-\frac{1}{2}c}^{\frac{1}{2}c} F(x_i+\xi)\,d\xi.$$

Since $|\,\xi\,|$ is small, then, by Taylor's expansion,

$$F(x_i+\xi)=\sum_{p=0}\frac{\xi^p}{p!}F^{(p)}(x_i); \tag{4}$$

hence
$$cy_i=N\left[\sum_{p=0}\frac{\xi^{p+1}}{(p+1)!}F^{(p)}(x_i)\right]_{-\frac{1}{2}c}^{\frac{1}{2}c}.$$

If p is odd, the corresponding terms on the right are zero. Then

$$cy_i=2N\Sigma\frac{(c/2)^{p+1}}{(p+1)!}F^{(p)}(x_i),$$

in which p takes even values, including zero. Thus for $p=0$, 2 and 4 we obtain

$$y_i = N\left[F(x_i) + \frac{c^2}{24}F''(x_i) + \frac{c^4}{1920}F^{\text{iv}}(x_i)\right].$$

This expression is correct to $O(c^5)$.

Hence, (3) becomes

$$m_r = \Sigma x_i^r F(x_i) + \frac{c^2}{24}\Sigma x_i^r F''(x_i) + \frac{c^4}{1920}\Sigma x_i^r F^{\text{iv}}(x_i).$$

We introduce the approximation

$$\Sigma x_i^r F^{(p)}(x_i) = \int_h^k x^r F^{(p)}(x)\,dx; \tag{5}$$

then $$m_r = \int x^r F(x)\,dx + \frac{c^2}{24}\int x^r F''(x)\,dx + \frac{c^4}{1920}\int x^r F^{\text{iv}}(x)\,dx, \tag{6}$$

the limits for the integrals being h and k.

By (2), the first term is μ_r. Now

$$\int x^r F''(x)\,dx = [x^r F'(x)] - r\int x^{r-1}F'(x)\,dx$$

$$= [x^r F'(x) - r x^{r-1}F(x)] + r(r-1)\int x^{r-2}F(x)\,dx. \tag{7}$$

Similarly,

$$\int x^r F^{\text{iv}}(x)\,dx = [x^r F'''(x) - r x^{r-1}F''(x)$$

$$+ r(r-1)\,x^{r-2}F'(x) - r(r-1)\,(r-2)x^{r-3}F(x)]$$

$$+ r\,.\,r-1\,.\,r-2\,.\,r-3\int x^{r-4}F(x)\,dx. \tag{8}$$

We shall assume for the present that h and k are finite. The conditions implied in applying Taylor's expansion in (4) require that the derivatives of $F(x)$ are finite and continuous in the range. We assume further that, for relevant values of p, $F^{(p)}(x)=0$ at each end of the range or, in other words, that the curve has close contact with the x-axis at $x=h$ and $x=k$. Accordingly, the integrated terms in (7) and (8) vanish; also, by means of (2), the integrals on the right-hand sides of (7) and (8) are expressed in terms of μ_{r-2} and μ_{r-4} respectively. We then have, from (6),

$$m_r = \mu_r + r(r-1)\frac{c^2}{24}\mu_{r-2} + r\,.\,r-1\,.\,r-2\,.\,r-3\,\frac{c^4}{1920}\mu_{r-4}. \tag{9}$$

We then have—putting $r=0, 1, \ldots, 4$ in succession, and remembering that by definition $m_0=1$—the following results:

$$m_0=\mu_0=1, \quad m_1=\mu_1=0,$$

$$m_2=\mu_2+\frac{c^2}{12}, \quad m_3=\mu_3,$$

$$m_4=\mu_4+\frac{c^2}{2}\mu_2+\frac{c^4}{80}.$$

From these, remembering that c is a small quantity, we obtain by a familiar reversion process

$$\left.\begin{aligned} &\mu_0=m_0=1, \quad \mu_1=m_1=0, \\ &\mu_2=m_2-\tfrac{1}{12}c^2, \quad \mu_3=m_3, \\ &\mu_4=m_4-\tfrac{1}{2}c^2m_2+\tfrac{7}{240}c^4. \end{aligned}\right\} \qquad (10)$$

If required, we obtain from (9)

$$\mu_5=m_5-\tfrac{5}{6}c^2m_3. \qquad (11)$$

The true principal moments, μ_r, are thus given by (10) and (11) in terms of the principal moments, m_r, calculated for the given statistical distribution; the terms involving c are Sheppard's corrections.

It is to be remembered that up to the present we have assumed that the range is finite, and it has been shown that, if Sheppard's corrections are applicable, then the curve must have close contact with the x-axis at the ends of the range.

Consider now the case when one or both limits are infinite; the vanishing of the integrated terms in (7) and (8) requires that

$$\lim_{|x|\to\infty}\{x^r F^{(p)}(x)\}=0;$$

this implies that the curve has contact of a very high order with the x-axis at one or both ends of the range.

These severe restrictions limit the applicability of Sheppard's corrections to distributions which have the properties mentioned and, in particular, to distributions which have the characteristics of the normal function such as the Gram-Charlier series described in § 7·05.

For many of Pearson's curves discussed in §§ 7·15–7·19 the restrictions clearly are not satisfied, and in such cases Sheppard's corrections must *not* be applied.

The legitimate application of Sheppard's corrections to a frequency distribution was illustrated in § 1·17.

8·04. The correction of an observed frequency distribution

It is assumed that the observational material relating to a characteristic, x, is presented in the form of a frequency polygon with a suitable class interval, and further that a smooth curve $y=v(x)$ has been drawn to give the best representation of the frequency distribution. A typical example concerns the counts of stars according to magnitude in a particular region of the sky, the class interval being, say, $0^{m}{\cdot}2$. The measures of magnitude are, of course, subject to errors, and, if these errors were supposed to be removed, we should obtain the true distribution represented by the curve $y=u(x)$.

In the sequel it is assumed that the function $u(x)$ representing the true distribution of a characteristic x and the function $v(x)$ representing the observed distribution are continuous functions with continuous derivatives within the range of x given by the statistics.

Let t denote the true value of the characteristic and x its measured value; the error, ϵ, of the latter is then given by $\epsilon=x-t$ or

$$x=t+\epsilon. \tag{1}$$

Let $\phi(\epsilon)$ be the function associated with the law of errors so that the relative frequency of errors between ϵ and $\epsilon+d\epsilon$ is $\phi(\epsilon)\,d\epsilon$.

The frequency of true values of the characteristic between t and $t+dt$ is $u(t)\,dt$, and of these the frequency with errors between ϵ and $\epsilon+d\epsilon$ is

$$u(t)\,\phi(\epsilon)\,dt\,d\epsilon.$$

Change the variables t, ϵ to x, ϵ; then, since by (1),

$$dt\,d\epsilon\equiv\frac{\partial(x-\epsilon,\epsilon)}{\partial(x,\epsilon)}\,dx\,d\epsilon=dx\,d\epsilon,$$

the frequency of the observed characteristic between x and $x+dx$ and with errors lying between ϵ and $\epsilon+d\epsilon$ is

$$u(x-\epsilon)\,\phi(\epsilon)\,dx\,d\epsilon. \tag{2}$$

The total frequency, $v(x)\,dx$, with observed characteristics between x and $x+dx$, will then be obtained by summing the expression (2) for all values of ϵ between $-\infty$ and $+\infty$. Hence

$$v(x)=\int_{-\infty}^{\infty}u(x-\epsilon)\,\phi(\epsilon)\,d\epsilon. \tag{3}$$

This is an integral equation in which the functions $v(x)$ and $\phi(\epsilon)$ are supposed to be known and from which the function $u(x)$ is to be derived.

We now assume that $\phi(\epsilon)$ is the function associated with the Gaussian law of errors so that

$$\phi(\epsilon)=\frac{h}{\sqrt{\pi}}e^{-h^2\epsilon^2},$$

the standard error, μ, being given by

$$\mu^2=\frac{1}{2h^2}. \tag{4}$$

From (3),
$$v(x)=\frac{h}{\sqrt{\pi}}\int_{-\infty}^{\infty}u(x-\epsilon)\,e^{-h^2\epsilon^2}\,d\epsilon. \tag{5}$$

Expand $u(x-\epsilon)$ by Taylor's theorem; then, $\frac{d^r u(x)}{dx^r}$ being denoted by $u_r(x)$, or, more simply, by u_r, (5) becomes

$$v(x)=\frac{h}{\sqrt{\pi}}\sum_{r=0}\frac{(-1)^r u_r}{r!}\int_{-\infty}^{\infty}\epsilon^r e^{-h^2\epsilon^2}\,d\epsilon$$
$$=\frac{1}{\sqrt{\pi}}\sum_0\frac{(-1)^r u_r}{h^r r!}\int_{-\infty}^{\infty}\xi^r e^{-\xi^2}\,d\xi.$$

If r is odd, the integral is zero. If r is even $(=2m)$, the integral is $\Gamma(m+\frac{1}{2})$ or $\frac{\sqrt{\pi}\,(2m)!}{2^{2m}m!}$. Hence

$$v(x)=\sum_0\frac{u_{2m}}{h^{2m}m!\,2^{2m}},$$

or, by means of (4),
$$v(x)=\sum_0\frac{u_{2m}}{m!}\left(\frac{\mu^2}{2}\right)^m,$$

in which m takes the values 0, 1, 2, 3, ...

The first few terms are

$$v(x)=u(x)+\frac{\mu^2}{2}u_2(x)+\frac{\mu^4}{4.2!}u_4(x)+\frac{\mu^6}{8.3!}u_6(x)+\dots. \tag{6}$$

Since μ is presumed to be a small quantity, the series in (6) can be inverted by a process of successive approximations. We can, however, derive the required results more rapidly by the use of the operator $D\ (\equiv d/dx)$, for (6) can be written as

$$v(x)=e^{\frac{1}{2}\mu^2D^2}u(x)$$

from which
$$u(x)=e^{-\frac{1}{2}\mu^2D^2}v(x)$$

or
$$u(x)=v(x)-\frac{\mu^2}{2}v_2(x)+\left(\frac{\mu^2}{2}\right)^2\frac{v_4(x)}{2!}-\dots; \tag{7}$$

the general term is $$(-1)^r \left(\frac{\mu^2}{2}\right)^r \frac{v_{2r}(x)}{r!}.$$

The formula (7) was first given by Eddington†; his proof is given in the next section.

8·05. Eddington's solution of the integral equation

The proof depends on the use of the operator D. Now

$$u(x-\epsilon) = u(x) - \epsilon Du(x) + \frac{\epsilon^2}{2!} D^2 u(x) - \ldots$$

by Taylor's theorem; then

$$u(x-\epsilon) = e^{-\epsilon D} u(x),$$

and consequently, by (5) of the previous section,

$$v(x) = \frac{h}{\sqrt{\pi}} \int_{-\infty}^{\infty} d\epsilon\, e^{-h^2\epsilon^2 - \epsilon D} u(x). \tag{1}$$

Now $$\int_{-\infty}^{\infty} e^{-h^2\epsilon^2 - a\epsilon} d\epsilon = \frac{\sqrt{\pi}}{h} e^{a^2/4h^2},$$

so that, if D is treated as a constant in (1),

$$v(x) = e^{D^2/4h^2} u(x) = e^{\frac{1}{2}\mu^2 D^2} u(x);$$

then $$u(x) = e^{-\frac{1}{2}\mu^2 D^2} v(x),$$

which leads to the series given by 8·04 (7).

8·06. Solution in terms of differences

We shall suppose that from the smoothed observed curve the values of $v(x)$ are obtained for values of the characteric

$$\ldots,\quad x-3c,\quad x-2c,\quad x-c,\quad x,\quad x+c,\quad x+2c,\quad \ldots,$$

where c is a small class interval.

We denote the second, fourth, ... differences by $\Delta_2, \Delta_4, \ldots$. Then, to $O(c^4)$,

$$\Delta_2 \equiv v(x+c) + v(x-c) - 2v(x)$$
$$= c^2 v_2(x) + \tfrac{1}{12} c^4 v_4(x),$$

and $$\Delta_4 \equiv v(x+2c) - 4v(x+c) + 6v(x) - 4v(x-c) + v(x-2c)$$
$$= c^4 v_4(x).$$

Hence $$v_2(x) = \frac{12\Delta_2 - \Delta_4}{12c^2} \quad \text{and} \quad v_4(x) = \frac{\Delta_4}{c^4}.$$

† A. S. Eddington, *Mon. Not. R. Astr. Soc.* **73** (1913), 359.

The first three terms of 8·04 (7) are then

$$u(x) = v(x) - \frac{(12\Delta_2 - \Delta_4)}{24}\left(\frac{\mu}{c}\right)^2 + \frac{\Delta_4}{8}\left(\frac{\mu}{c}\right)^4. \tag{1}$$

This approximation will suffice for most purposes. If the fourth differences are not well-defined, it will be sufficient to take

$$u(x) = v(x) - \tfrac{1}{2}\Delta_2\left(\frac{\mu}{c}\right)^2. \tag{2}$$

On the other hand, Eddington† considered a particular problem: 'The effect of Red-shift on the magnitudes of Nebulae', in which it was essential and practicable to include terms up to the eighth difference in the expansion for $u(x)$.

8·07. Illustration of the correction of a frequency distribution

The method is illustrated with reference to the number of stars with measured parallaxes‡ grouped between the values 0″·010 and 0″·020, 0″·020 and 0″·030, etc. (Table 18). It is convenient to take 0″·001 as the unit so that the class interval, c, is 10 and the abscissae of the points P_i on the frequency polygon are 15, 25, ..., denoted by x_i in the first column; the corresponding frequencies are in the second column. From a smoothed curve, which is subject to some extent to individual judgement, the smoothed frequencies are obtained; these are shown, to the nearest unit, as $v(x)$ in the third column. The first and second differences, Δ_1 and Δ_2, are in the next two columns.

The probable error is stated to be 0″·0096 or 9·6 in the unit adopted; the standard error, μ, is then given by $\mu = 9{\cdot}6/0{\cdot}6745$ or by $\mu^2 = 202{\cdot}5$;

Table 18. *Counts of parallaxes*

x_i	f_i	$v(x)$	Δ_1	Δ_2	$u(x)$
15	106	106			
			−18		
25	88	88		+3	85
			−15		
35	75	73		0	73
			−15		
45	62	58*		+1	57
			−14		
55	33	44		+2	42
			−12		
65	30	32		+2	30
			−10		
75	27	22		+2	20
			−8		
85	12	14		+3	11
			−5		
95	10	9			

† A. S. Eddington, *Mon. Not. R. Astr. Soc.* **97** (1937), 156.
‡ J. J. Nassau, *Mon. Not. R. Astr. Soc.* **88** (1928), 584.

hence $\frac{1}{2}(\mu/c)^2 = 1{\cdot}01$. The corrected distribution, $u(x)$, is found from the formula 8·06 (2) involving second differences only, namely,

$$u(x) = v(x) - \tfrac{1}{2}\Delta_2\left(\frac{\mu}{c}\right)^2.$$

The values of $u(x)$, for x between 25 and 85, are shown in the last column; these may be regarded as the values of the distribution, to the approximation adopted, freed from accidental errors.

8·08. 'Improved' value of a measure

From 8·04 (2) the frequency associated with the observed characteristic between x and $x+dx$ and with errors between ϵ and $\epsilon+d\epsilon$ is, on writing $\phi(\epsilon)$ in terms of the normal function,

$$\frac{h}{\sqrt{\pi}}\, u(x-\epsilon)\, e^{-h^2\epsilon^2}\, dx\, d\epsilon.$$

Let $\bar{\epsilon}$ denote the mean error corresponding to the observed characteristic between x and $x+dx$. Then

$$\bar{\epsilon}\int_{-\infty}^{\infty} u(x-\epsilon)\, e^{-h^2\epsilon^2}\, d\epsilon = \int_{-\infty}^{\infty} \epsilon u(x-\epsilon)\, e^{-h^2\epsilon^2}\, d\epsilon,$$

or, by 8·04 (5)—the integral on the right being integrated by parts—

$$\bar{\epsilon}\,\frac{\sqrt{\pi}}{h}\, v(x) = -\frac{1}{2h^2}\left[u(x-\epsilon)\, e^{-h^2\epsilon^2}\right]_{-\infty}^{\infty} + \frac{1}{2h^2}\int_{-\infty}^{\infty} e^{-h^2\epsilon^2}\frac{d}{d\epsilon}\{u(x-\epsilon)\}\, d\epsilon.$$

The integrated part on the right vanishes at each limit. Also,

$$\frac{d}{d\epsilon}\{u(x-\epsilon)\} = -\frac{d}{dx}\{u(x-\epsilon)\};$$

hence
$$\bar{\epsilon}v(x) = -\frac{1}{2h\sqrt{\pi}}\frac{d}{dx}\int_{-\infty}^{\infty} u(x-\epsilon)\, e^{-h^2\epsilon^2}\, d\epsilon,$$

or, by 8·04 (5),
$$\bar{\epsilon} = -\frac{1}{2h^2}\frac{v'(x)}{v(x)} = -\mu^2\frac{v'(x)}{v(x)}, \tag{1}$$

a result† depending on the smoothed distribution, the gradient for which, at a selected value of x, is usually obtained with considerable accuracy.

Eddington‡ defines the 'improved' value, ξ, of a measure, x, by

$$\xi = x - \bar{\epsilon}; \tag{2}$$

† Due originally to A. S. Eddington and first published by F. W. Dyson, *Mon. Not. R. Astr. Soc.* **86** (1926), 686.

‡ A. S. Eddington, *Mon. Not. R. Astr. Soc.* 100 (1940), 359.

this is the best value to be associated with an individual measure. Eddington has pointed out the fallaciousness of applying the correction, $\bar{\epsilon}$, to the smoothed distribution, $v(x)$—as has sometimes been done—so as to obtain an 'improved' distribution; it has to be emphasized that the correction has validity only when it is applied to an individual measure.

As an illustration we find the 'improved' value of a parallax measure between $0''{\cdot}040$ and $0''{\cdot}050$. If y denotes an ordinate of the smoothed curve, (1) can be written as

$$\bar{\epsilon} = -\mu^2 \frac{1}{y} \frac{\Delta y}{\Delta x}. \tag{3}$$

Here, $y = 58$ (indicated by an asterisk in Table 18), and, approximately,

$$\frac{\Delta y}{\Delta x} = -\frac{(73-44)}{2c} = -\frac{29}{20}.$$

Further, $\mu^2 = 202{\cdot}5$. Hence, from (3), $\bar{\epsilon} = +5$ or $+0''{\cdot}005$.

If the parallax, p, for a particular star in the range concerned is $0''{\cdot}044$, the 'improved' value is given, from (2), by $p = 0''{\cdot}044 - \bar{\epsilon}$ or $0''{\cdot}039$.

8·09. The integral equation

From 8·04 (3) the fundamental integral equation in general form is

$$v(x) = \int_{-\infty}^{\infty} u(x-y)\,\phi(y)\,dy, \tag{1}$$

in which y replaces ϵ.

This can be written in an alternative form. Put $\xi = x - y$; then, for a given x, $d\xi = -dy$ and (1) becomes

$$v(x) = -\int_{\infty}^{-\infty} u(\xi)\,\phi(x-\xi)\,d\xi,$$

or, replacing ξ by y,

$$v(x) = \int_{-\infty}^{\infty} u(y)\,\phi(x-y)\,dy. \tag{2}$$

In general, if the functions $v(x)$ and $\phi(y)$ are known, the function $u(x)$ can be found explicitly in terms of an integral involving Fourier transforms which will be considered in the next section.

8·10. Fourier transforms

We begin with the well-known double integral

$$f(x) = \frac{1}{\pi}\int_0^{\infty} dt \int_{\infty}^{\infty} f(y)\cos t(x-y)\,dy, \tag{1}$$

where $f(x)$ is a function satisfying Dirichlet's conditions and, in particular, $\int_{-\infty}^{\infty} f(x)\,dx$ is absolutely convergent; the conditions mentioned are satisfied by the kind of functions which we are likely to enounter in the present connexion.

From (1),

$$f(x)=\frac{1}{\pi}\int_0^{\infty}\cos tx\left\{\int_{-\infty}^{\infty} f(y)\cos ty\,dy\right\}dt$$
$$+\frac{1}{\pi}\int_0^{\infty}\sin tx\left\{\int_{-\infty}^{\infty} f(y)\sin ty\,dy\right\}dt.$$

Denote the integrals within the brackets by $P(t)$ and $Q(t)$ respectively, with the obvious properties: $P(-t)=P(t)$ and $Q(-t)=-Q(t)$. Then

$$f(x)=\frac{1}{2\pi}\int_0^{\infty} P(t)\{e^{itx}+e^{-itx}\}\,dt-\frac{i}{2\pi}\int_0^{\infty} Q(t)\{e^{itx}-e^{-itx}\}\,dt. \tag{2}$$

Set
$$F(t)\equiv P(t)+iQ(t)=\int_{-\infty}^{\infty} f(y)\,e^{ity}\,dy \tag{3}$$

and
$$F(-t)\equiv P(t)-iQ(t)=\int_{-\infty}^{\infty} f(y)\,e^{-ity}\,dy.$$

Then (2) becomes

$$f(x)=\frac{1}{2\pi}\int_0^{\infty} F(t)\,e^{-itx}\,dt+\frac{1}{2\pi}\int_0^{\infty} F(-t)\,e^{itx}\,dt.$$

If t is written for $-t$ in the second integral it becomes

$$\int_{-\infty}^{0} F(t)\,e^{-itx}\,dt.$$

Hence
$$f(x)=\frac{1}{2\pi}\int_{-\infty}^{\infty} F(t)\,e^{-ixt}\,dt. \tag{4}$$

Now put $F(t)=\surd(2\pi)f^*(t)$. Then, from (3) and (4), we have, on writing x for y in the former,

$$f^*(t)=\frac{1}{\surd(2\pi)}\int_{-\infty}^{\infty} f(x)\,e^{itx}\,dx \tag{5}$$

and
$$f(x)=\frac{1}{\surd(2\pi)}\int_{-\infty}^{\infty} f^*(t)\,e^{-ixt}\,dt. \tag{6}$$

These formulae are Fourier transforms and the functions $f(x)$ and $f^*(x)$ are said to be conjugate.

8·11. Solution of the integral equation

From 8·09 (1), the integral equation is

$$v(x)=\int_{-\infty}^{\infty} u(x-y)\,\phi(y)\,dy. \tag{1}$$

Multiply throughout by e^{itx} and integrate from $-\infty$ to ∞ with respect to x; then

$$\int_{-\infty}^{\infty} v(x)\,e^{itx}\,dx=\int_{-\infty}^{\infty} \phi(y)\,e^{ity}\left\{\int_{-\infty}^{\infty} u(x-y)\,e^{it(x-y)}\,dx\right\}dy.$$

For a given y the integral within the brackets on the right-hand side is $\int_{-\infty}^{\infty} u(\xi)\,e^{it\xi}\,d\xi$ or $\sqrt{(2\pi)}\,u^*(t)$, by 8·10 (5). Then also, by 8·10 (5),

$$\sqrt{(2\pi)}\,v^*(t)=2\pi\phi^*(t)\,u^*(t);$$

hence
$$u^*(t)=\frac{1}{\sqrt{(2\pi)}}\frac{v^*(t)}{\phi^*(t)},$$

and consequently, by 8·10 (6),

$$u(x)=\frac{1}{2\pi}\int_{-\infty}^{\infty}\frac{v^*(t)}{\phi^*(t)}\,e^{-ixt}\,dt. \tag{2}$$

In general the integrand is a complex function and the solution, $u(x)$, which is a real function, is the real part of the integral.

In particular, if $v(x)$ and $\phi(y)$ are even functions of x and y, then, for example,

$$v^*(t)=\frac{1}{\sqrt{(2\pi)}}\int_{-\infty}^{\infty} v(x)\cos tx\,dx+\frac{i}{\sqrt{(2\pi)}}\int_{-\infty}^{\infty} v(x)\sin tx\,dx;$$

the second integral vanishes and hence

$$v^*(t)=\frac{1}{\sqrt{(2\pi)}}\int_{-\infty}^{\infty} v(x)\cos tx\,dx. \tag{3}$$

Similarly,
$$\phi^*(t)=\frac{1}{\sqrt{(2\pi)}}\int_{-\infty}^{\infty} \phi(x)\cos tx\,dx.$$

Thus, $v^*(t)$ and $\phi^*(t)$ are even functions of t and accordingly

$$u(x)=\frac{1}{2\pi}\int_{-\infty}^{\infty}\frac{v^*(t)}{\phi^*(t)}\cos xt\,dt.$$

8·12. Application to normal functions

In the integral equation

$$v(x)=\int_{-\infty}^{\infty} u(x-y)\,\phi(y)\,dy,$$

we assume that, in terms of the means x_0 and y_0 and the standard deviations β and σ,

$$v(x)=\frac{1}{\beta\sqrt{(2\pi)}}\,e^{-(x-x_0)^2/2\beta^2} \tag{1}$$

and

$$\phi(y)=\frac{1}{\sigma\sqrt{(2\pi)}}\,e^{-(y-y_0)^2/2\sigma^2}. \tag{2}$$

Then, by 8·10 (5),

$$v^*(t)=\frac{1}{2\pi\beta}\int_{-\infty}^{\infty} e^{-(x-x_0)^2/2\beta^2+itx}\,dx$$

$$=\frac{e^{itx_0}}{2\pi\beta}\int_{-\infty}^{\infty} e^{-\xi^2/2\beta^2}\cos t\xi\,d\xi,$$

by 8·11 (3) and setting $\xi=x-x_0$.

Now, by a well-known result,

$$\int_{-\infty}^{\infty} e^{-a^2\xi^2}\cos b\xi\,d\xi=\frac{\sqrt{\pi}}{a}\,e^{-b^2/4a^2}. \tag{3}$$

Hence

$$v^*(t)=\frac{1}{\sqrt{(2\pi)}}\,e^{-\frac{1}{2}\beta^2t^2+itx_0}.$$

Similarly,

$$\phi^*(t)=\frac{1}{\sqrt{(2\pi)}}\,e^{-\frac{1}{2}\sigma^2t^2+ity_0}. \tag{4}$$

Then, by 8·11 (2),

$$u(x)=\frac{1}{2\pi}\int_{-\infty}^{\infty} e^{-\frac{1}{2}\alpha^2t^2+it(x_0-y_0-x)}\,dt$$

$$=\frac{1}{2\pi}\int_{-\infty}^{\infty} e^{-\frac{1}{2}\alpha^2t^2}\cos t(x-x_1)\,dt,$$

where

$$\alpha^2=\beta^2-\sigma^2 \quad\text{and}\quad x_1=x_0-y_0;$$

it is to be noted that it is assumed that $\beta>\sigma$. Hence, by (3),

$$u(x)=\frac{1}{\alpha\sqrt{(2\pi)}}\,e^{-(x-x_1)^2/2\alpha^2}.$$

Thus, $u(x)$ is a normal function with $x_1\equiv x_0-y_0$ as the mean and α as the standard deviation.

If $\phi(\epsilon)$ is the error function, with $\sigma\equiv\mu$ as the standard deviation, given by

$$\phi(\epsilon)=\frac{1}{\mu\sqrt{(2\pi)}}\,e^{-\epsilon^2/2\mu^2},$$

then $u(x)$ is given by

$$u(x)=\frac{1}{\alpha\sqrt{(2\pi)}}e^{-(x-x_0)^2/2\alpha^2},$$

where

$$\alpha^2=\beta^2-\mu^2.$$

Thus the effect of removing errors from the observed function $v(x)$ is to express the true function $u(x)$ as a normal function with the same mean as $v(x)$ and with a standard deviation, α, less than that of $v(x)$.

8·13. Application to the Gram-Charlier series

We shall suppose that the observed function $v(x)$ has been obtained as a Gram-Charlier series and given, with a change of notation in 7·08 (5), by

$$v(x)=\frac{1}{\beta\sqrt{(2\pi)}}\left[\Sigma\frac{(-1)^n A_n}{\beta^n n!}H_n(\xi)\right]e^{-(x-x_0)^2/2\beta^2},$$

where $\xi=(x-x_0)/\beta$ and the coefficients A_n have been found in terms of the principal moments as in 7·08 (3); also, $A_0=1$, $A_1=A_2=0$.

Let

$$B_n=\frac{(-1)^n A_n}{\beta^n n!}. \tag{1}$$

By 8·10 (5),

$$\begin{aligned}v^*(t)&=\frac{1}{2\pi\beta}e^{itx_0}\Sigma\int_{-\infty}^{\infty}B_n H_n(\xi)\,e^{-\frac{1}{2}\xi^2+it(x-x_0)}\,dx\\&=\frac{1}{2\pi}e^{-\frac{1}{2}\beta^2t^2+itx_0}\Sigma B_n\int_{-\infty}^{\infty}H_n(\xi)\,e^{-\frac{1}{2}(\xi-T)^2}\,d\xi,\end{aligned} \tag{2}$$

where $T=i\beta t$.

Let I_n denote the integral in (2); then, on putting $\eta=\xi-T$, we have

$$I_n=\int_{-\infty}^{\infty}H_n(\xi)\,e^{-\frac{1}{2}\eta^2}\,d\eta.$$

Now, by 7·07 (4),

$$H_n=\xi H_{n-1}-\frac{dH_{n-1}}{d\xi};$$

hence

$$\begin{aligned}I_n&=\int_{-\infty}^{\infty}\left[\eta H_{n-1}-\frac{dH_{n-1}}{d\eta}\right]e^{-\frac{1}{2}\eta^2}\,d\eta+TI_{n-1}\\&=-[H_{n-1}\,e^{-\frac{1}{2}\eta^2}]_{-\infty}^{\infty}+TI_{n-1}.\end{aligned}$$

The integrated part vanishes; consequently

$$I_n=TI_{n-1}.$$

Hence

$$I_n=T^nI_0=\sqrt{(2\pi)}\,(i\beta t)^n$$

and $$v^*(t)=\frac{1}{\sqrt{(2\pi)}}\,e^{-\frac{1}{2}\beta^2t^2+itx_0}\,\Sigma B_n(i\beta t)^n.$$

Also, when $\phi(y)$ is given by 8·12 (2), then, by 8·12 (4),

$$\phi^*(t)=\frac{1}{\sqrt{(2\pi)}}\,e^{-\frac{1}{2}\sigma^2t^2+ity_0}.$$

By 8·11 (2), the solution of the integral equation is

$$u(x)=\frac{1}{2\pi}\,\Sigma B_n(i\beta)^n\int_{-\infty}^{\infty}t^n\,e^{-\frac{1}{2}\alpha^2t^2-it(x-x_0+y_0)}\,dt,$$

where $\alpha^2=\beta^2-\sigma^2$.

Put $$X=x-x_0+y_0 \quad\text{and}\quad S=-iX/\alpha^2;$$

then $$u(x)=\frac{1}{2\pi}\,e^{-\frac{1}{2}X^2/\alpha^2}\,\Sigma B_n(i\beta)^n\int t^n\,e^{-\frac{1}{2}\alpha^2(t-S)^2}\,dt.$$

Now write $$F(x)=\frac{1}{\alpha\sqrt{(2\pi)}}\,e^{-\frac{1}{2}X^2/\alpha^2} \quad\text{and}\quad \eta=\alpha(t-S);$$

then $$u(x)=\frac{F(x)}{\sqrt{(2\pi)}}\,\Sigma B_n(i\beta)^n\int_{-\infty}^{\infty}t^n\,e^{-\frac{1}{2}\eta^2}\,d\eta. \tag{3}$$

Let K_n denote the integral; then

$$K_n=\int_{-\infty}^{\infty}\left(\frac{\eta}{\alpha}+S\right)t^{n-1}\,e^{-\frac{1}{2}\eta^2}\,d\eta$$

$$=-\frac{1}{\alpha}\,[t^{n-1}\,e^{-\frac{1}{2}\eta^2}]_{-\infty}^{\infty}+\frac{n-1}{\alpha}\int_{-\infty}^{\infty}t^{n-2}\frac{dt}{d\eta}\,d\eta+SK_{n-1}.$$

The integrated part vanishes; hence

$$K_n=SK_{n-1}+\frac{n-1}{\alpha^2}K_{n-2}. \tag{4}$$

Now $$S=-\frac{iX}{\alpha^2}=\frac{1}{i\alpha}\tau,$$

where $$\tau=X/\alpha.$$

Also, $K_0=\sqrt{(2\pi)}$ and, by (4), $K_1=\dfrac{\tau}{i\alpha}K_0$.

Again, from (4),

$$K_2=K_0\left[\frac{\tau^2}{(i\alpha)^2}+\frac{1}{\alpha^2}\right]=\frac{K_0}{(i\alpha)^2}[\tau^2-1].$$

These results can be written as

$$K_1/K_0=\frac{1}{i\alpha}H_1(\tau),\quad K_2/K_0=\frac{1}{(i\alpha)^2}H_2(\tau);$$

also, they suggest that $K_r/K_0=\frac{1}{(i\alpha)^r}H_r(\tau).$ (5)

If this formula holds for $r=n-1$ and $r=n-2$, we shall show that it holds for $r=n$. From (4) and (5),

$$K_n/K_0=\frac{\tau}{i\alpha}\left[\frac{1}{(i\alpha)^{n-1}}H_{n-1}(\tau)\right]+\frac{n-1}{\alpha^2}\frac{1}{(i\alpha)^{n-2}}H_{n-2}(\tau)$$

$$=\frac{1}{(i\alpha)^n}[\tau H_{n-1}(\tau)-(n-1)H_{n-2}(\tau)]$$

$$=\frac{1}{(i\alpha)^n}H_n(\tau),\quad \text{by 7·07 (3)}.$$

The formula (5) holds for $r=1$ and $r=2$; hence it holds for $r=3$ and consequently for all positive integral values of n. Hence, from (1),

$$B_n(i\beta)^n K_n=\sqrt{(2\pi)}\left\{(-1)^n\frac{A_n}{\alpha^n n!}\right\}H_n(\tau)$$

and (3) becomes

$$u(x)=\frac{1}{\alpha\sqrt{(2\pi)}}e^{-\frac{1}{2}\tau^2}\left[1+\sum_3\frac{(-1)^n A_n}{\alpha^n n!}H_n(\tau)\right],\tag{6}$$

where $$\tau\equiv\frac{X}{\alpha}=\frac{1}{\alpha}(x-x_0+y_0).$$

The formula (6) shows that $u(x)$ is a Gram-Charlier series with the same coefficients, A_n, as in the series for $v(x)$, the normal function associated with $u(x)$ being

$$F(x)=\frac{1}{\alpha\sqrt{(2\pi)}}e^{-(x-x_0+y_0)^2/2\alpha^2}.$$

If $\phi(y)$ is the error function $\frac{1}{\sigma\sqrt{(2\pi)}}e^{-y^2/2\sigma^2}$, then $y_0=0$ and the function $F(x)$ is a normal function with x_0 as the mean and α as the standard deviation, α being given by $\alpha^2=\beta^2-\sigma^2$.

8·14. Correction of vectors

In Fig. 30 OT is a true vector of magnitude r and components (ξ,η) with respect to the reference axes OA and OB; OS is the observed vector of magnitude ρ and with components (x,y) with respect to the axes OX (drawn through OT) and OY.

We shall assume that the law of errors is the same for the two components ξ and η and consequently for the components x and y; and that it is given for each, independently, by the normal function $\phi(\epsilon)=\frac{h}{\sqrt{\pi}}e^{-h^2\epsilon^2}$. Now $TR\equiv x-r$ is the error for the x-component of OS and $RS\equiv y$ is the error for the y-component. The probability that an error will lie between $x-r$ and $x-r+dx$ is $\phi(x-r)\,dx$, and the probability that an error will lie between y and $y+dy$ is $\phi(y)\,dy$. Hence the probability that the observed vector will have components between $x-r$ and $x-r+dx$ and between y and $y+dy$ is

$$\frac{h^2}{\pi}e^{-h^2\{(x-r)^2+y^2\}}\,dx\,dy.$$

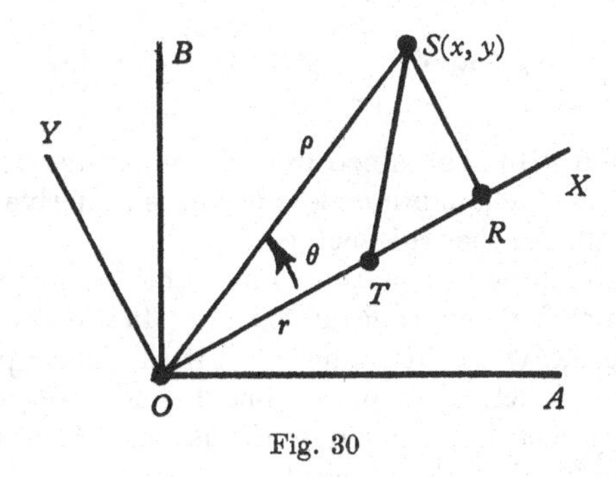

Fig. 30

Now, if θ is the angle SOT, then

$$(x-r)^2+y^2\equiv ST^2=r^2+\rho^2-2\rho r\cos\theta.$$

Hence the probability that the observed vector will have a magnitude between ρ and $\rho+d\rho$ and lie in the sector $d\theta$ is

$$\frac{h^2}{\pi}e^{-h^2(r^2+\rho^2-2\rho r\cos\theta)}\,\rho\,d\rho\,d\theta.$$

Further, the probability $\zeta d\rho$ that an assigned true vector OT will give rise to an observed vector with magnitude between ρ and $\rho+d\rho$, and irrespective of its direction, is obtained by summing the previous expression for all values of θ from O to 2π. Thus

$$\zeta d\rho=\frac{2h^2}{\pi}e^{-h^2(r^2+\rho^2)}\,\rho\,d\rho\int_0^{\pi}e^{2h^2\rho r\cos\theta}\,d\theta$$

$$=2h^2e^{-h^2(r^2+\rho^2)}\,\rho\,d\rho\,I_0(2h^2\rho r), \tag{1}$$

where $I_0(t)$ is the modified Bessel function of zero order defined by

$$I_0(t) = \sum_0^\infty \frac{(\frac{1}{2}t)^{2n}}{(n!)^2}. \tag{2}$$

Suppose now that N, the number of vectors, is large. For a given value of r the frequency of true vectors between r and $r+dr$ is $Nf(r)\,dr$ and the corresponding frequency of observed vectors with magnitudes between ρ and $\rho+d\rho$ is $Nf(r)\,dr\,\zeta\,d\rho$. If $NF(\rho)\,d\rho$ is the total number of vectors observed to have magnitudes between ρ and $\rho+d\rho$ for *all* true vectors OT irrespective of their directions, then

$$\begin{aligned} F(\rho) &= \int_0^\infty \zeta f(r)\,dr \\ &= 2h^2\rho\, e^{-h^2\rho^2} \int_0^\infty I_0(2h^2\rho r) f(r)\, e^{-h^2r^2}\,dr. \end{aligned} \tag{3}$$

If the function $F(\rho)$ is obtained from the measures, then (3) is an integral equation which, in principle, enables us to derive the relative frequency function of the true vectors.

In general, the integral equation is insoluble. If, however, $f(r)$ is given a particular form which leads to the evaluation of the integral in (3), then we derive the function $F(\rho)$; if the latter proves to be substantially the function obtained from the measures, then we can infer, conversely, that $f(r)$ is given by the assumed form.

If $f(r) = Ar\,e^{-\beta^2r^2}$, then the integral in (3) can be evaluated. Since $f(r)$ is the relative frequency then

$$\int_0^\infty f(r)\,dr = 1 = A\int_0^\infty r\,e^{-\beta^2r^2}\,dr,$$

from which $A = 2\beta^2$. The assumed function is then

$$f(r) = 2\beta^2 r\, e^{-\beta^2r^2}\,dr. \tag{4}$$

Then, if we write

$$H^2 = h^2 + \beta^2 \tag{5}$$

and put y for $2h^2\rho$ in I_0, (3) becomes

$$F(\rho) = 4h^2\beta^2\rho\, e^{-h^2\rho^2} \int_0^\infty rI_0(yr)\, e^{-H^2r^2}\,dr. \tag{6}$$

Let K denote the integral; then, by (2),

$$K = \int_0^\infty \sum_0^\infty \frac{y^{2n} r^{2n+1}\, e^{-H^2r^2}\,dr}{2^{2n}(n!)^2}.$$

Now
$$\int_0^\infty r^{2n+1}e^{-H^2r^2}dr = \frac{1}{H^{2n+2}}\int_0^\infty t^{2n+1}e^{-t^2}dt$$
$$= \frac{1}{H^{2n+2}}\tfrac{1}{2}n!.$$

Hence
$$K = \frac{1}{2H^2}\sum_0^\infty \left(\frac{y^2}{4H^2}\right)^n \frac{1}{n!} = \frac{1}{2H^2}e^{y^2/4H^2}. \tag{7}$$

Hence, replacing y by $2h^2\rho$, we find that (6) becomes
$$F(\rho) = \frac{2h^2\beta^2}{H^2}\rho\, e^{-h^2\beta^2\rho^2/H^2},$$
or, if $k = h\beta/H$,
$$F(\rho) = 2k^2\rho\, e^{-k^2\rho^2}. \tag{8}$$

Thus, $F(\rho)$ is of the same analytical form as $f(r)$.

Conversely, if $F(\rho)$ is the observed function given by (8) in which k is supposed known, then the function $f(r)$ of the true vectors is given by
$$f(r) = 2\beta^2 r\, e^{-\beta^2 r^2},$$
where, from (5) and since $k = h\beta/H$,
$$\beta^2 = \frac{h^2k^2}{h^2-k^2}, \tag{9}$$
in which it is necessary to assume that $h > k$.

The results† of this section are due to J. C. Kapteyn and P. J. van Rhijn.

8·15. Correction of the mean of the observed magnitudes of vectors

In some problems the mean of the observed magnitudes of vectors, denoted by $\bar{\rho}$, is the starting point of a specific investigation. The value of $\bar{\rho}$ is subject, of course, to error, and the object of this section is to obtain the relevant correction so as to obtain the true mean, $\bar{r}$, to be used subsequently.

It is assumed that the observed relative distribution can be expressed as
$$F(\rho) = 2A_1k_1^2\rho\, e^{-k_1^2\rho^2} + 2A_2k_2^2\rho\, e^{-k_2^2\rho^2} + \dots, \tag{1}$$
in which the A's and k's are known.

Accordingly, the function representing the true relative distribution is given by
$$f(r) = 2A_1\beta_1^2 r\, e^{-\beta_1^2 r^2} + 2A_2\beta_2^2 r\, e^{-\beta_2^2 r_2^2} + \dots,$$
where
$$\beta_1^2 = h^2k_1^2/(h^2-k_1^2), \text{ etc.,} \quad \text{with } h > k_n.$$

† *Groningen Publ.* no. 30 (1920), p. 44.

Then $$\bar{r}=\int_0^\infty rf(r)\,dr.$$

For a particular term in $f(r)$, with constants A and β,

$$\int_0^\infty rf(r)\,dr=2A\beta^2\int_0^\infty r^2 e^{-\beta^2r^2}\,dr=\frac{A\sqrt{\pi}}{2\beta}.$$

Hence $$\bar{r}=\frac{\sqrt{\pi}}{2}\left\{\frac{A_1}{\beta_1}+\frac{A_2}{\beta_2}+\ldots\right\}.$$

Similarly, $$\bar{\rho}=\frac{\sqrt{\pi}}{2}\left\{\frac{A_1}{k_1}+\frac{A_2}{k_2}+\ldots\right\}.$$

Hence, if c denotes the correction to be applied to $\bar{\rho}$, then

$$\bar{r}=\bar{\rho}+c,$$

where $$c=-\frac{\sqrt{\pi}}{2}(A_1\alpha_1+A_2\alpha_2+\ldots) \tag{2}$$

and $$\alpha_1=\frac{1}{k_1}-\frac{1}{\beta_1},\quad \text{etc.} \tag{3}$$

Now, in general, if β and k are any pair of β_n and k_n, then

$$\beta^2=h^2k^2/(h^2-k^2)$$

and hence $$\beta^2-k^2=k^4/(h^2-k^2)$$

from which $\beta>k$, since $h>k$; thus, the α's are all positive.

Further, since $\int_0^\infty F(\rho)\,d\rho=1$, we have from (1),

$$A_1+A_2+\ldots=1.$$

Frequently, one term in (1) is predominant, say, the first, in which case c is negative so that the true mean is less than the observed mean.

8·16. Relative frequency of the magnitudes of vectors exceeding a given value

As regards the observed vectors, if $S(\rho)$ denotes the relative frequency when the magnitude exceeds ρ, then

$$S(\rho)=\int_\rho^\infty F(\rho)\,d\rho=\Sigma 2A_n k_n^2\int_\rho^\infty \rho\, e^{-k_n^2\rho^2}d\rho$$
$$=\Sigma A_n e^{-k_n^2\rho^2}.$$

Similarly, if $S(r)$ denotes the relative frequency when the true magnitude exceeds r, then

$$S(r) = \Sigma A_n e^{-\beta_n^2 r^2}.$$

The function $e^{-k^2x^2}$ is tabulated in Appendix 3. The statistics can then be interpreted very readily in terms of this function.

If, for example, the function $F(\rho)$ consists of one term only, then $A=1$; it is then comparatively easy, by taking several values of k and using graphical methods, to derive that value of k which best represents the function $S(\rho)$.

Further, if μ and r_1 are the standard error and probable error associated with the modulus, h, then

$$\frac{1}{h^2} = 2\mu^2 = 2\left(\frac{r_1}{0{\cdot}6745}\right)^2 = 4{\cdot}395 r_1^2.$$

Then β is obtained from the formula 8·14 (9) so that

$$\frac{1}{\beta^2} = \frac{1}{k^2} - 4{\cdot}395 r_1^2. \tag{1}$$

8·17. Example of the correction of the mean observed magnitude of vectors

In Table 19, relating to the total proper motions† of 100 stars, the second column gives the number of observed magnitudes between 0″·0 and 0″·05, between 0″·05 and 0″·10, etc.; one star, with $\rho = 3''{\cdot}0$, is counted as ½ for the range 0″·25–0″·30 and ½ for the range 0″·30–0″·35. The third column gives the number $N(\rho)$ of magnitudes

Table 19. *Total proper motions (centennial)*

ρ	Number observed	$N(\rho)$	$S(\rho)$ (observed)	$S(\rho)$ (calculated)
0″·0		100	1·00	1·00
0″·5	5	95	0·95	0·94
1″·0	12	83	0·83	0·79
1″·5	26	57	0·57	0·59
2″·0	16	41	0·41	0·40
2″·5	18	23	0·23	0·24
3″·0	8½	14½	0·14	0·13
3″·5	7½	7	0·07	0·06
4″·0	2	5	0·05	0·03
>4″·0	5			

† W. M. Smart, *Stellar Dynamics* (Cambridge University Press, 1938), p. 216.

exceeding 0″·0, 0″·5, etc., and, with $N=100$, the fourth column gives the corresponding values of $S(\rho)$, as defined in the previous section. It is found by plotting the function $e^{-k^2\rho^2}$ (Appendix 3) for several values of k that a good representation of the values of $S(\rho)$ in the fourth column is obtained with $k=0{\cdot}48$. With this value of k the corresponding calculated values of $S(\rho)$ are given in the last column, the details of which are in good agreement with the observed values $S(\rho)$.

The probable error, r_1, is 0″·4; hence, with $k=0{\cdot}48$, we find from 8·16 (1) that $\beta=0{\cdot}525$. From 8·15 (2) and (3), the correction, c, is given by

$$c=-\frac{\sqrt{\pi}}{2}\left(\frac{1}{k}-\frac{1}{\beta}\right)$$

or $c=-0''{\cdot}158$. If, then, $\bar{\rho}$ is the mean value of the total proper motions, the corrected mean, $\bar{r}$, is given by

$$\bar{r}=\bar{\rho}-0''{\cdot}158.$$

8·18. 'Improved' value of the magnitude of an observed vector

From 8·14 (1), $\zeta d\rho$ is the probability that the observed magnitude of a vector lies between ρ and $\rho+d\rho$. Also, $Nf(r)\,dr$ is the frequency of true vectors with magnitudes between r and $r+dr$. Hence the frequency of vectors with measured magnitudes between ρ and $\rho+d\rho$ and with true magnitudes between r and $r+dr$ is

$$N\zeta f(r)\,dr\,d\rho.$$

This expression can be equally well regarded as the frequency of vectors with true magnitudes between r and $r+dr$ and with measured magnitudes between ρ and $\rho+d\rho$. Hence, for a given value of ρ, the mean, $\bar{r}$, of the true magnitudes is given by

$$\bar{r}=\frac{\int_0^\infty r\zeta f(r)\,dr}{\int_0^\infty \zeta f(r)\,dr},$$

or, by 8·14 (1),

$$\bar{r}=\frac{\int_0^\infty r\,e^{-h^2r^2}f(r)\,I_0(yr)\,dr}{\int_0^\infty e^{-h^2r^2}f(r)\,I_0(yr)\,dr},$$

where $y=2h^2\rho$.

We assume in the first instance, that

$$f(r)=2\beta^2 r\, e^{-\beta^2 r^2}. \tag{1}$$

Then, by 8·14 (5), $$\bar{r}=\frac{\int_0^\infty r^2 e^{-H^2r^2} I_0(yr)\,dr}{\int_0^\infty r\, e^{-H^2r^2} I_0(yr)\,dr}\equiv\frac{M}{K}.$$

The denominator is given by 8·14 (7), namely,

$$K=\frac{1}{2H^2}e^{y^2/4H^2}. \tag{2}$$

The numerator, M, is expressed as

$$M=\sum_0^\infty\left(\frac{y^2}{4}\right)^n\frac{1}{(n!)^2}\int_0^\infty r^{2n+2}e^{-H^2r^2}\,dr$$

$$=\frac{1}{H^3}\sum\left(\frac{y^2}{4H^2}\right)^n\frac{1}{(n!)^2}\int_0^\infty t^{2n+2}e^{-t^2}\,dt.$$

The integral is $\qquad \frac{1}{2}\Gamma(n+\frac{3}{2})=\frac{(2n+2)!}{2^{2n+2}(n+1)!}\frac{\sqrt{\pi}}{2},$

which can be written as

$$\frac{\sqrt{\pi}}{4}\frac{2n+1}{2^n}[2n-1.2n-3.\dots.1].$$

Hence $$M=\frac{\sqrt{\pi}}{4H^3}\sum_0^\infty\left(\frac{y^2}{4H^2}\right)^n\frac{1+2n}{n!}\left[\frac{1.3.5.\dots.2n-1}{2.4.6.\dots.2n}\right].$$

But, by a well-known result,

$$\frac{1.3.5.\dots.2n-1}{2.4.6.\dots.2n}=\frac{1}{\pi}\int_0^\pi\cos^{2n}\theta\,d\theta;$$

hence, if $$q=y^2/(4H^2)=2b, \tag{3}$$

$$M=\frac{1}{4H^3\sqrt{\pi}}\int_0^\pi\left\{\sum_0^\infty\frac{(q\cos^2\theta)^n}{n!}+2q\cos^2\theta\sum_1^\infty\frac{(q\cos^2\theta)^{n-1}}{(n-1)!}\right\}d\theta$$

$$=\frac{1}{4H^3\sqrt{\pi}}\int_0^\pi(1+2q\cos^2\theta)\,e^{q\cos^2\theta}\,d\theta$$

$$=\frac{e^b}{4H^3\sqrt{\pi}}\int_0^\pi(1+2b+2b\cos2\theta)\,e^{b\cos2\theta}\,d\theta.$$

Now, $I_0(b)$ can be expressed as an integral as follows:

$$\int_0^\pi e^{b\cos 2\theta}\,d\theta = \pi I_0(b),$$

from which $$\int_0^\pi \cos 2\theta\, e^{b\cos 2\theta}\,d\theta = \pi \frac{d}{db} I_0(b) = \pi I_1(b),$$

where $I_0(b)$ and $I_1(b)$ are the modified Bessel functions of order zero and one respectively. Hence

$$M = \frac{e^b\sqrt{\pi}}{4H^3}[(1+2b)\,I_0(b)+2bI_1(b)].$$

From (2) and (3), $K = \dfrac{1}{2H^2}e^{2b}$; hence

$$\bar{r} = \frac{\sqrt{\pi}}{2H}\psi(b), \tag{4}$$

where $$\psi(b) = e^{-b}[(1+2b)\,I_0(b)+2bI_1(b)]. \tag{5}$$

The values of the functions $e^{-x}I_0(x)$ and $e^{-x}I_1(x)$ have been tabulated† and hence the values of the function $\psi(b)$ can be easily found‡ for various values of b.

It is more convenient to express $\bar{r}$, in (4), in terms of a function $G(q)$ defined by

$$G(q) = \frac{\sqrt{\pi}}{2}\psi(b),$$

in which $$q = 2b = y^2/4H^2$$

and $$y = 2h^2\rho, \quad H^2 = h^2+\beta^2,$$

so that $$q = \frac{h^4\rho^2}{H^2}.$$

Then $$\bar{r} = \frac{1}{H}G(q).$$

The function $G(q)$ was introduced§ by J. C. Kapteyn and P. J. van Rhijn and evaluated by an approximation process.

The values of $G(q)$ are given in Appendix 2; these values have been obtained by means of the accurate formula (5).

† G. N. Watson, *Theory of Bessel Functions* (Cambridge University Press, 1922), pp. 698–713.

‡ See, for example, W. M. Smart, *Stellar Dynamics* (Cambridge University Press, 1938), p. 43.

§ *Groningen Publ.* no. 30 (1920), p. 63.

If $F(\rho)$ is expressed, in general, as

$$F(\rho)=2A_1k_1^2\rho\,e^{-k_1^2\rho^2}+2A_2k_2^2\rho\,e^{-k_2^2\rho^2}+\ldots,$$

the formula for $\bar{r}$ becomes

$$\bar{r}=\frac{\Sigma\dfrac{A_n\beta_n^2}{H_n^3}e^{q_n}G(q_n)}{\Sigma\dfrac{A_n\beta_n^2}{H_n^2}e^{q_n}}, \tag{6}$$

where
$$\left.\begin{aligned} H_n^2=h^2+\beta_n^2,\quad q_n=h^4\rho^2/H_n^2,\\ \beta_n^2=\frac{h^2k_n^2}{h^2-k_n^2}.\end{aligned}\right\} \tag{7}$$

It is assumed that the observed distribution leads to the evaluation of the A's and k's. The constants in (6) are then found by means of (7).

For a given observed value, ρ, the value, $\bar{r}$, is the 'improved' value.

For example, taking the distribution in Table 19, we shall find the 'improved' value corresponding to a vector for which $\rho=3''\cdot0$.

Now, $h^2=\dfrac{1}{2\mu^2}=\dfrac{(0\cdot6745)^2}{2r_1^2}=1\cdot422$, with $r_1=0''\cdot4$. Also, we found $\beta=0\cdot525$; hence, $H^2\equiv h^2+\beta^2=1\cdot698$. Then $q=1\cdot191\rho^2$, so that, for $\rho=3''\cdot0$, $q=10\cdot72$.

From Appendix 2, $G(q)=3\cdot351$; hence

$$\bar{r}\equiv\frac{1}{H}G(q)=2''\cdot57.$$

Thus, the 'improved' value of an individual observed value $3''\cdot0$ is $2''\cdot57$.

8·19. Correction of a function of vectors by the method of operators

As before, $u(x,y)$ and $v(x,y)$ are the relative frequencies of the true and observed vectors; we assume that the modulus, h, of the error function ϕ is the same for each coordinate. By analogy with 8·04 (3) the integral equation is

$$v(x,y)=\int_{-\infty}^{\infty}\int_{-\infty}^{\infty}u(x-\alpha,y-\beta)\,\phi(\alpha)\,\phi(\beta)\,d\alpha\,d\beta,$$

where α and β denote the errors in the two coordinates.

By Taylor's theorem

$$u(x-\alpha,y-\beta)=e^{-\alpha D-\beta D_1}u(x,y),$$

where D and D_1 are the operators $\partial/\partial x$ and $\partial/\partial y$ respectively. Then

$$v(x,y)=\frac{h^2}{\pi}\int_{-\infty}^{\infty} e^{-h^2\alpha^2-\alpha D}\,d\alpha \int_{-\infty}^{\infty} e^{-h^2\beta^2-\beta D_1}\,d\beta\, u(x,y),$$

from which, D and D_1 being regarded as constants,

$$v(x,y)=e^{(D^2+D_1^2)/4h^2}\,u(x,y)$$

so that
$$u(x,y)=\exp\left[-\left(\frac{\partial^2}{\partial x^2}+\frac{\partial^2}{\partial y^2}\right)\Big/4h^2\right]v(x,y).$$

Since the standard error, μ, is given by $\mu^2=1/2h^2$, we have, to $O(\mu^3)$,

$$u(x,y)=v(x,y)-\frac{\mu^2}{2}\left\{\frac{\partial^2 v}{\partial x^2}+\frac{\partial^2 v}{\partial y^2}\right\}.$$

If c is the class interval in each coordinate then

$$u(x,y)=v(x,y)-\frac{\mu^2}{2c^2}\{\Delta_2(x)+\Delta_2(y)\}, \tag{1}$$

where $\Delta_2(x)$ denotes the second difference in the values of x with an assigned value of y, and $\Delta_2(y)$ is defined in a similar manner.

In terms of polar coordinates r and θ,

$$\frac{\partial^2 v}{\partial x^2}+\frac{\partial^2 v}{\partial y^2}=\frac{1}{r}\frac{\partial v}{\partial r}+\frac{\partial^2 v}{\partial r^2}+\frac{1}{r^2}\frac{\partial^2 v}{\partial \theta^2}.$$

If the function $v(r,\theta)$ is tabulated at intervals c in r and at intervals b in $r\theta$, then, in terms of first and second order differences and to $O(\mu^3)$,

$$u(r,\theta)=v(r,\theta)-\frac{\mu^2}{2}\left\{\frac{1}{cr}\Delta_1(r)+\frac{\Delta_2(r)}{c^2}+\frac{\Delta_2(\theta)}{b^2}\right\}. \tag{2}$$

It is clear that the use of the formulae (1) or (2) requires the statistics of a very large number of vectors; otherwise, the necessary smoothing of the distribution $v(x,y)$ for (i) y constant and (ii) x constant or alternatively, the smoothing of the distribution $v(r,\theta)$ for (i) θ constant and (ii) r constant, cannot be achieved with any measure of reliability.

CHAPTER 9

CORRELATION

9·01. Introduction

Consider two distinct attributes, age and height, of a large number of individuals; as a matter of simple observation men of 50 years of age are on the average taller than boys of 10 years of age, and the latter are taller on the average than boys of 4 years of age. There is evidently a somewhat general statistical relationship between age, A, and height, H. This is not a *precise* relationship, for, if it were so, there would be a definite connexion between age and height which could be expressed by an equation of the form: $H = f(A)$.

Again, for example, of a class of thirty schoolboys studying Latin and Greek it might reasonably be expected by classical masters—in the light of previous experience—that, in general, proficiency in one subject would be related in a fairly close way to proficiency in the other, and that a poor performance in one language would be accompanied by a poor performance in the other; in other words, there would be a general relationship between the performances in the two languages, but it would not necessarily be a precise one, expressible in terms of examination marks, in the form $L = f(G)$.

We take another example from physics. The relation between the pressure, p, of a gas, its volume, v, and its absolute temperature, T, is given by $pv = RT$, where R is a constant or, in the more convenient form for statistical purposes,

$$x + y = z + C,$$

where $x = \log p$, $y = \log v$ and $z = \log T$.

If T is constant, we have Boyle's law (1662) and, if p is constant, we have Charles's law (1787).

If we *suppose*, by way of illustration, that Boyle's observations had been made without any reference to temperature—which, we might suppose, had any value between 273° (absolute) and 300° in his series of measures—the relation $pv =$ constant, or

$$x + y = c, \tag{1}$$

would not then be an *exact* formula connecting p and v or x and y, although the aggregate of points (x, y) would be found to be scattered fairly closely about some such line as PQ in Fig. 31. Such a diagram is called a *scatter diagram*.

Although x and y are not related by the precise form (1), it would be evident that there is *some* connexion, or statistical relationship, between x and y suggesting that, possibly, some other factor was operating to produce the dispersion of the points in the diagram and that further experiments with stricter controls might reveal it.

In such cases as we have considered we express this general relationship by saying that there exists a *correlation* between the two attributes or characteristics concerned.

On the other hand, there is no conceivable relationship between the weekly number of babies born in the Soviet Union during the four months of May to August 1956 and the weekly aggregate of runs scored by the Australian cricketers in first-class matches in England during the same four months. We say in this case that there is no correlation between the two sets of statistics.

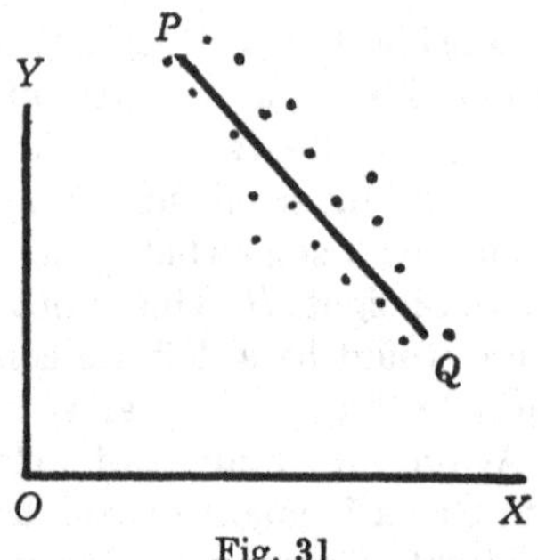

Fig. 31

Our hypothetical example on Boyle's law, which, of course, has no close relevance to the historical sequence of the discoveries mentioned, illustrates the importance of following up, with further investigations, a series of observations involving two entities and revealing a high degree of correlation. In this connexion an illustration from astronomy stresses this particular point. About 1914 it was established that there was a close correlation between the luminosity, L, of the stars of the 'main sequence' with spectral type which we now associate with effective temperature, T. It was noticed that the hottest stars had the largest masses, M and, progressively, the cooler stars had the smaller masses. In a sense this was a challenge to theoretical investigators of the internal constitution of the stars culminating in Eddington's discovery in 1924 of the mass-luminosity relationship, substantially of the form $L=f(M,T)$, which revolutionized the astronomical thought of the time.

The application of correlation methods has thus, initially, at least, two possible aims: first, to discover the degree of dependance of one variable on another and secondly, in many instances, to suggest the desirability for further observational and theoretical investigations along new directions.

9·02. Covariance

Suppose that of N individuals the number (or frequency) f_i have the pair of measurable attributes, such as age and height, which we denote by x_i and y_i; such a distribution is called a *bivariate distribution*. Then, as usual, if $\bar{x}$ and $\bar{y}$ denote the means of the two attributes,

$$N\bar{x}=\Sigma f_i x_i, \quad N\bar{y}=\Sigma f_i y_i. \tag{1}$$

The variances, σ_x^2 and σ_y^2, corresponding to the distributions of x and y respectively, are, by 1·06 (5), given by

$$N\sigma_x^2=\Sigma f_i(x_i-\bar{x})^2, \quad N\sigma_y^2=\Sigma f_i(y_i-\bar{y})^2. \tag{2}$$

Also, by 1·06 (14), if $\mu_2(a)$ denotes the second-order moment, with reference to the values of x_i about the line $x=a$, then

$$\mu_2(a)=\sigma_x^2+(\bar{x}-a)^2. \tag{3}$$

In particular,

$$\mu_2(0)=\sigma_x^2+\bar{x}^2. \tag{4}$$

We have formulae, similar to (3) and (4) for the second-order moment of the values of y_i about the line $y=b$; thus, if ν is defined for y in the same way as μ is defined for x,

$$\nu_2(b)=\sigma_y^2+(\bar{y}-b)^2,$$

and, in particular,

$$\nu_2(0)=\sigma_y^2+\bar{y}^2.$$

We now introduce the *product moment* about (a,b), denoted by $\mu_{11}(a,b)$ and defined by

$$N\mu_{11}(a,b)=\Sigma f_i(x_i-a)(y_i-b), \tag{5}$$

or, on expanding the right-hand side and using (1),

$$N\mu_{11}(a,b)=\Sigma f_i x_i y_i+N(ab-a\bar{y}-b\bar{x}). \tag{6}$$

Here, $\frac{1}{N}\Sigma f_i x_i y_i\equiv\mu_{11}(0,0)$ is the product moment about the origin.

The product moment with respect to $(\bar{x},\bar{y})$, that is, $\mu_{11}(\bar{x},\bar{y})$, is given the special name of *covariance*, usually denoted by p; thus, from (6), with $a=\bar{x}$ and $b=\bar{y}$,

$$Np=\Sigma f_i x_i y_i-N\bar{x}\bar{y}. \tag{7}$$

Hence, from (6) and (7),

$$N[p-\mu_{11}(a,b)]=N(a\bar{y}+b\bar{x}-\bar{x}\bar{y}-ab)$$

or

$$p=\mu_{11}(a,b)-(\bar{x}-a)(\bar{y}-b). \tag{8}$$

In particular,

$$p=\mu_{11}(0,0)-\bar{x}\bar{y}. \tag{9}$$

In calculations it is usually much more convenient to derive the value of the product moment $\mu_{11}(a, b)$, where a and b are suitably chosen and then to obtain the covariance, p, by means of (8).

As we shall see later the covariance is one of the factors which determine the degree of correlation between the two characteristics under investigation.

9·03. Lines of regression

The information relating to pairs of characteristics (x_i, y_i) can be conveniently displayed in a *scatter diagram*, as illustrated in Fig. 32, in which the related pair of characteristics (x_i, y_i) is represented by a dot at P_i with reference to axes OX, OY. In many instances it is

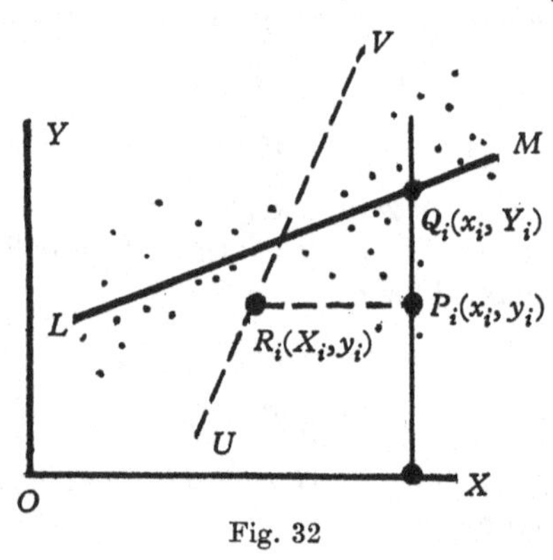

Fig. 32

found that the assembly of dots suggests that the points are distributed, more or less closely, about a straight line. The problem is to determine the equation of the line which best represents the statistics.

Let

$$y = mx + c \tag{1}$$

be the equation of a line LM, in which m and c are two disposable constants to be evaluated according to some criterion. Let the ordinate through $P_i(x_i, y_i)$ meet LM at $Q_i(x_i, Y_i)$; then $Y_i = mx_i + c$ and $P_iQ_i = mx_i + c - y_i$. We further suppose, in general, that the frequency of pairs (x_i, y_i) is f_i.

One criterion for determining the line LM which best represents the statistics is that

$$E \equiv \Sigma f_i . P_iQ_i^2 \equiv \Sigma f_i(mx_i + c - y_i)^2 \tag{2}$$

should be a minimum; the conditions are

$$\frac{\partial E}{\partial c} = 0 \quad \text{and} \quad \frac{\partial E}{\partial m} = 0.$$

The first condition, N being the total frequency, gives

$$\Sigma f_i(mx_i + c - y_i) \equiv Nm\bar{x} + cN - N\bar{y} = 0$$

or
$$\bar{y}=m\bar{x}+c. \tag{3}$$

From (1) and (3),
$$y-\bar{y}=m(x-\bar{x}), \tag{4}$$

which shows that the required line passes through the centroid $(\bar{x},\bar{y})$ of the distribution.

The second condition for a minimum, namely, $\partial E/\partial m=0$, gives

$$\Sigma f_i x_i(mx_i+c-y_i)=0$$

or
$$\Sigma f_i x_i y_i=m\Sigma f_i x_i^2+c\Sigma f_i x_i$$

or
$$\mu_{11}(0,0)=m\mu_2(0)+c\bar{x}$$
$$=m\mu_2(0)+\bar{x}\bar{y}-m\bar{x}^2, \quad \text{by (3)}.$$

Now, by 9·02 (4),
$$\mu_2(0)=\sigma_x^2+\bar{x}^2,$$

and by 9·02 (9),
$$\mu_{11}(0,0)=p+\bar{x}\bar{y};$$

hence
$$p=m\sigma_x^2. \tag{5}$$

The equation of the line, given by (4), becomes

$$y-\bar{y}=\frac{p}{\sigma_x^2}(x-\bar{x}).$$

The *correlation coefficient*, r, is defined by

$$r=\frac{p}{\sigma_x\sigma_y}. \tag{6}$$

The equation of the line is then written as

$$\frac{y-\bar{y}}{\sigma_y}=r\frac{x-\bar{x}}{\sigma_x}. \tag{7}$$

This line is called the *line of regression of y on x*.

It is to be observed that LM, the line of regression of y on x, is derived according to the criterion that $\Sigma f_i P_i Q_i^2$ should be a minimum.

An alternative criterion to give the line best in accordance with the statistics—represented by UV in Fig. 32—is that $\Sigma f_i R_i P_i^2$ should be a minimum, R_i being the point of intersection of UV and the parallel to OX through P_i.

If the equation of UV is written as

$$x=My+C,$$

the coordinates of R_i are (X_i, y_i), where $X_i=My_i+C$ and the condition becomes that

$$E\equiv\Sigma f_i(My_i+C-x_i)^2$$

should be a minimum. By analogy with the previous work it is readily seen that UV is

$$\frac{x-\bar{x}}{\sigma_x}=r\frac{y-\bar{y}}{\sigma_y}, \tag{8}$$

where r is defined as in (6).

This line is called the *line of regression of x on y.*

Referred to the centroid as origin, the line of regression of y on x is

$$y=\frac{r\sigma_y}{\sigma_x}x,$$

and the line of regression of x on y is

$$x=\frac{r\sigma_x}{\sigma_y}y.$$

The two lines of regression are identical only if $r^2=1$. By (6), the sign of r is that of p which may be positive or negative.

9·04. The correlation coefficient

Let S_y^2 denote the mean of the squared deviations $P_iQ_i^2$ in Fig. 32, so that

$$\begin{aligned} NS_y^2&=\Sigma f_i(mx_i+c-y_i)^2\\ &=\Sigma f_i(mx_i+\bar{y}-m\bar{x}-y_i)^2, \quad \text{by 9·03 (3)},\\ &=\Sigma f_i[m(x_i-\bar{x})-(y_i-\bar{y})]^2. \end{aligned}$$

Hence $$S_y^2=m^2\sigma_x^2+\sigma_y^2-2mp.$$

Now, by (5) and (6) of the previous section,

$$m^2\sigma_x^2=p^2/\sigma_x^2=r^2\sigma_y^2$$

and $$2mp=2p^2/\sigma_x^2=2r^2\sigma_y^2;$$

hence $$S_y^2=\sigma_y^2(1-r^2).$$

This equation shows that $|r|\leqslant 1$. Further, if $|r|=1$, then

$$S_y^2\equiv\frac{1}{N}\Sigma f_iP_iQ_i^2=0,$$

from which it follows that all the points P_i lie on a straight line; in this case the characteristics x and y are functionally related and the correlation is then perfect. When $|r|<1$, the degree of correlation is expressed in terms of this value of $|r|$. If $r=0$, the lines of regression are lines through the centroid parallel to the axes OX and OY; in this case there is no correlation. The correlation may be regarded as *significant* if $|r|$ is not less than 0·5 and as highly significant if $|r|$ is 0·8.

9·05. Example on the calculation of the correlation coefficient

About a quarter of a century ago when a sufficient number of observations of eclipsing binary stars became available, Lundmark† drew attention to the high degree of correlation between the masses and the radii of the components of these systems.

In Table 20 are given the masses and radii of the more massive components of thirty-three well-observed systems‡; the mass, M, and radius, R, are expressed in terms of the Sun's mass and radius as units.

To simplify the calculations we write

$$x = M - 6, \quad y = R - 4;$$

the statistics are then represented by the x and y columns, and the remaining three columns are formed. The means of the separate columns are shown at the feet of the columns; they are

$$\bar{x} = 0{\cdot}72, \quad \bar{y} = -0{\cdot}18, \quad \frac{1}{N}\Sigma x^2 = 87{\cdot}2, \quad \frac{1}{N}\Sigma y^2 = 19{\cdot}6$$

and
$$\frac{1}{N}\Sigma xy = 37{\cdot}8.$$

We then have
$$\sigma_x^2 = 87{\cdot}2 - \bar{x}^2 = 86{\cdot}7,$$
$$\sigma_y^2 = 19{\cdot}6 - \bar{y}^2 = 19{\cdot}6,$$
and, by 9·02 (9),
$$p = 37{\cdot}8 - \bar{x}\bar{y} = 37{\cdot}9.$$
From these,
$$\sigma_x^2 \sigma_y^2 = 1699{\cdot}3 = (41{\cdot}2)^2$$
and
$$r \equiv \frac{p}{\sigma_x \sigma_y} = \frac{37{\cdot}9}{41{\cdot}2},$$
that is,
$$r = 0{\cdot}92.$$

The large value of r shows the high degree of correlation existing between M and R.

The line of regression of y on x is readily found to be, from 9·03 (7),

$$y + 0{\cdot}18 = 0{\cdot}437(x - 0{\cdot}72).$$

Similarly, from 9·03 (8), the line of regression of x on y is

$$y + 0{\cdot}18 = 0{\cdot}517(x - 0{\cdot}72).$$

The angles of slope are 23°·6 and 27°·3 respectively; the lines pass through the centroid (0·72, −0·18) and are mutually inclined at the small angle of 3°·7.

† *Lund Observatory Circular*, no. 5 (1932), p. 105.

‡ C. Payne-Gaposchkin and S. Gaposchkin, *Variable Stars* (Harvard Observatory Monographs, no. 5 (Cambridge (Mass.), 1938), pp. 41, 43.

Table 20. *Masses and radii of the more massive components of binary stars*

	M	R	x $(M-6)$ +	x −	y $(R-4)$ +	y −	x^2	y^2	xy +	xy −
1	0·8	1·2	—	5·2	—	2·8	27·0	7·8	14·6	—
2	6·2	3·5	0·2	—	—	0·5	0·0	0·2	—	0·1
3	6·7	3·0	0·7	—	—	1·0	0·5	1·0	—	0·7
4	2·2	2·1	—	3·8	—	1·9	14·4	3·6	7·2	—
5	2·7	1·9	—	3·3	—	2·1	10·9	4·4	6·9	—
6	2·4	3·3	—	3·6	—	0·7	13·0	0·5	2·5	—
7	0·6	0·6	—	5·4	—	3·4	29·2	11·6	18·4	—
8	1·8	3·4	—	4·2	—	0·6	17·6	0·4	2·5	—
9	1·8	2·2	—	4·2	—	1·8	17·6	3·2	7·6	—
10	36·3	22·9	30·3	—	18·9	—	918·1	357·2	572·7	—
11	40·3	17·0	34·3	—	13·0	—	1176·5	169·0	445·9	—
12	14·6	5·4	8·6	—	1·4	—	74·0	2·0	12·0	—
13	7·0	3·0	1·0	—	—	1·0	1·0	1·0	—	1·0
14	17·6	4·4	11·6	—	0·4	—	134·6	0·2	4·6	—
15	1·0	1·5	—	5·0	—	2·5	25·0	6·2	12·5	—
16	3·0	2·5	—	3·0	—	1·5	9·0	2·3	4·5	—
17	0·6	1·0	—	5·4	—	3·0	29·2	9·0	16·2	—
18	1·6	3·0	—	4·4	—	1·0	19·4	1·0	4·4	—
19	2·2	1·8	—	3·8	—	2·2	14·4	4·8	8·4	—
20	2·1	1·7	—	3·9	—	2·3	15·2	5·3	9·0	—
21	7·5	4·5	1·5	—	0·5	—	2·2	0·2	0·7	—
22	5·0	6·3	—	1·0	2·3	—	1·0	5·3	—	2·3
23	1·4	1·8	—	4·6	—	2·2	21·2	4·8	10·1	—
24	2·0	1·3	—	4·0	—	2·7	16·0	7·3	10·8	—
25	5·4	2·8	—	0·6	—	1·2	0·4	1·4	0·7	—
26	1·4	1·3	—	4·6	—	2·7	21·2	7·3	12·4	—
27	5·2	3·7	—	0·8	—	0·3	0·6	0·1	0·2	—
28	21·2	6·9	15·2	—	2·9	—	231·0	8·4	44·1	—
29	6·7	2·0	0·7	—	—	2·0	0·5	4·0	—	1·4
30	0·8	0·8	—	5·2	—	3·2	27·0	10·2	16·6	—
31	3·0	1·3	—	3·0	—	2·7	9·0	7·3	8·1	—
32	5·3	3·3	—	0·7	—	0·7	0·5	0·5	0·5	—
33	5·3	4·5	—	0·7	0·5	—	0·5	0·2	—	0·3
			104·1	**80·4**	**39·9**	**46·0**	**2877·7**	**647·7**	**1254·1**	**5·8**
		Means		**+0·72**		**−0·18**	**+87·2**	**+19·6**	**+37·8**	

9·06. Contingency tables

If the number of individuals in a bivariate distribution is large, the method of deriving the correlation coefficient illustrated in Table 20 involves an immense amount of calculation which can be greatly lightened, without any substantial loss of accuracy, by means of a *contingency table*. This can be formed most conveniently from a scatter diagram. Suppose that the statistics can be reduced in terms of x and y, which may be positive or negative; by drawing the lines $x = \pm a, \pm 3a, \pm 5a, \ldots$ and $y = \pm b, \pm 3b, \pm 5b, \ldots$, we can then count the number of dots lying within rectangles each of area $4ab$; if one such number is f, we then assume that the distribution within the rectangle is equivalent to f dots at the centre of the rectangle; the various sums Σfx, Σfy, Σfx^2, Σfy^2 and Σfxy can then be readily obtained. The method is illustrated in the following example.

9·07. Example on the use of a contingency table

We consider the distribution of parallaxes, Π, and proper motions, μ, of stars in two antipodal regions of the sky. In general it would be expected that there would be some correlation between Π and μ for, other things being equal, the nearer the stars—that is, the larger the values of Π—the larger would be the values of μ.

The statistics† in the following tables refer to values of Π from $0''{\cdot}025$ upwards and for values of μ from $0''{\cdot}10$ upwards, the stars concerned lying in one region, defined by right ascensions $4\frac{1}{2}^{\text{h}}$ and $7\frac{1}{2}^{\text{h}}$ and declinations positive, and in the antipodal region.

The stars, 146 in number, for which $0''{\cdot}025 \leqslant \Pi \leqslant 0''{\cdot}105$ and $0''{\cdot}10 \leqslant \mu \leqslant 1''{\cdot}10$ are treated by means of the contingency table 21.

There are fifteen stars with larger values of Π and μ, and these were dealt with individually as in Table 20; only the results will be given later.

We take $\Pi = 0''{\cdot}05$ and $\mu = 0''{\cdot}45$ as convenient values of a and b and write

$$x = 100\Pi - 5, \quad y = 10\mu - 4{\cdot}5.$$

In Table 21 the class interval for Π is $0''{\cdot}1$, with the middle values, Π_m, equal to $0''{\cdot}3$, $0''{\cdot}4$, ..., that is, with unit class interval for x and middle values $-2, -1, 0, \ldots$.

Similarly, the class interval for μ is $0''{\cdot}1$, with middle values, μ_m, equal to $0''{\cdot}15$, $0''{\cdot}25$, ..., that is, with unit class interval for y and middle values $-3, -2, -1, \ldots$.

In the first of the two rows in Table 21 corresponding to each value of y, the number of stars in each rectangle of the scatter diagram is inserted; for example, corresponding to $x = -2$ and $y = -3$, the number of stars with Π between $0''{\cdot}025$ and $0''{\cdot}035$ (that is, with $\Pi_m = 0''{\cdot}03$ and with $x = -2$) and with μ between $0''{\cdot}10$ and $0''{\cdot}20$ (that is, with $\mu_m = 0''{\cdot}15$ and $y = -3$) is 22.

In the second-last column are found the numbers, f, of stars corresponding to each value of y and for all values of x; thus, there are thirty-four stars for which $y = -3$ for all values of x.

† F. Schlesinger, *General Catalogue of Stellar Parallaxes* (Yale University Observatory, 1935).

Table 21. *Contingency table for* 146 *stars*

μ_m	Π_m	0″·03	0″·04	0″·05	0″·06	0″·07	0″·08	0″·09	0″·10		
	$y \backslash x$	-2	-1	0	1	2	3	4	5	f	fxy
0″·15	-3	22	6	5	1	0	0	0	0	34	—
		132	18	0	-3	0	0	0	0	—	147
0″·25	-2	11	9	6	4	4	0	1	1	36	—
		44	18	0	-8	-16	0	-8	-10	—	20
0″·35	-1	7	7	4	2	2	1	1	0	24	—
		14	7	0	-2	-4	-3	-4	0	—	8
0″·45	0	4	5	2	3	1	2	0	0	17	—
		0	0	0	0	0	0	0	0	—	0
0″·55	1	3	3	1	3	1	1	1	0	13	—
		-6	-3	0	3	2	3	4	0	—	3
0″·65	2	2	0	2	2	0	0	1	0	7	—
		-8	0	0	4	0	0	8	0	—	4
0″·75	3	3	1	0	0	2	0	0	0	6	—
		-18	-3	0	0	12	0	0	0	—	-9
0″·85	4	0	1	1	1	1	0	0	1	5	—
		0	-4	0	4	8	0	0	20	—	28
0″·95	5	1	0	1	0	0	0	0	0	2	—
		-10	0	0	0	0	0	0	0	—	-10
1″·05	6	0	0	0	1	1	0	0	0	2	—
		0	0	0	6	12	0	0	0	—	18
	Sums	**53**	**32**	**22**	**17**	**12**	**4**	**4**	**2**	**146**	**209**

Similarly, corresponding to each value of x the number, f_1, of stars for all values of y is inserted at the foot of the corresponding column; thus, there are fifty-three stars for $x=-2$ for all values of y.

In the second row corresponding to each value of y in Table 21 are found the values of fxy for each value of x; for example, for $x=-2$, $y=-3$ and $f=22$, the product is 132. The first entry in the final column, namely, 147, is the sum, denoted simply by fxy, of all such products formed for all the values of x and for $y=-3$. The other entries in the last column are obtained in a similar way. From the sum at the foot of the last column we have, for the 146 stars of the table,

$$\Sigma fxy = 209. \tag{1}$$

In Table 22, f_1 is the number of stars, for all values of y, corresponding to each value of x; these are the numbers found at the bottom of the columns in Table 21. The products $f_1 x$ and $f_1 x^2$ are formed and summed.

Similarly, in Table 23, f is the number of stars, for all values of x, corresponding to each value of y; these are the numbers found in the second-last column of Table 21. The products fy and fy^2 are formed and summed.

Table 22

x	f_1	f_1x +	f_1x −	f_1x^2
−2	53	—	106	212
−1	32	—	32	32
0	22	0	—	0
1	17	17	—	17
2	12	24	—	48
3	4	12	—	36
4	4	16	—	64
5	2	10	—	50
	146	**−59**		**459**

Table 23

y	f	fy +	fy −	fy^2
−3	34	—	102	306
−2	36	—	72	144
−1	24	—	24	24
0	17	0	—	0
1	13	13	—	13
2	7	14	—	28
3	6	18	—	54
4	5	20	—	80
5	2	10	—	50
6	2	12	—	72
	146	**−111**		**771**

The results of Tables 22 and 23 are summarized as follows:

$$\Sigma f_1 x = -59, \quad \Sigma f y = -111,$$

$$\Sigma f_1 x^2 = 459, \quad \Sigma f y^2 = 771,$$

and, from (1), $$\Sigma fxy = 209.$$

The results for the fifteen stars outside the ranges of the contingency table are merely stated here; they are, with $f=f_1=1$,

$$\Sigma f_1 x = 141, \quad \Sigma f y = 101,$$

$$\Sigma f_1 x^2 = 1550, \quad \Sigma f y^2 = 1121,$$

$$\Sigma fxy = 1031.$$

Combining the above results with those for the 146 stars of the contingency table we have:

$$\Sigma f_1 x = 82, \quad \Sigma fy = -10,$$

$$\Sigma f_1 x^2 = 2009, \quad \Sigma fy^2 = 1892,$$

$$\Sigma fxy = 1240.$$

The total number, N, of stars is 161. Then,

$$161\bar{x} = 82, \quad 161\bar{y} = -10,$$

$$161\mu_1^2 = 2009, \quad 161\mu_2^2 = 1892,$$

$$161\mu_{11}(0,0) = 1240,$$

where $$N\mu_1^2 = \Sigma f_1 x^2, \quad N\mu_2^2 = \Sigma fy^2$$

and $$N\mu_{11}(0,0) = \Sigma fxy.$$

We obtain:

$$\bar{x} = 0{\cdot}51, \quad \bar{y} = -0{\cdot}06, \quad \mu_1^2 = 12{\cdot}48, \quad \mu_2^2 = 11{\cdot}75$$

and $\mu_{11}(0,0) = 7{\cdot}70$. With these values we find

$$\mu_x^2 \equiv \mu_1^2 - \bar{x}^2 = 12{\cdot}22, \quad \mu_y^2 \equiv \mu_2^2 - \bar{y}^2 = 11{\cdot}71$$

and $$p \equiv \mu_{11}(0,0) - \bar{x}\bar{y} = 7{\cdot}73.$$

Finally, $$r \equiv \frac{p}{\sigma_x \sigma_y} = 0{\cdot}646.$$

For the stars concerned there is thus a high degree of correlation between their parallaxes and proper motions.

9·08. Parameters for continuous bivariate distributions

We suppose that there is a very large number, N, of individuals each with an associated pair of characteristics, denoted in general by x and y and represented as in the scatter diagram. Let $\phi(x, y)$ denote a function such that the probability that a point in the x, y plane has coordinates between x and $x+dx$ and between y and $y+dy$ is

$$\phi(x, y)\, dx\, dy.$$

As in the case of the normal function it is stipulated that $\phi \to 0$ as $x \to \pm\infty$ and as $y \to \pm\infty$.

If dn denotes the probable number of points in the area $dx\,dy$, then

$$dn = N\phi(x, y)\, dx\, dy.$$

The mean, $\bar{x}$, of all the values of x is given by

$$\bar{x}=\frac{1}{N}\Sigma x\,dn=\int_{-\infty}^{\infty}\int_{-\infty}^{\infty}x\phi(x,y)\,dx\,dy.$$

Similarly,
$$\bar{y}=\int_{-\infty}^{\infty}\int_{-\infty}^{\infty}y\phi(x,y)\,dx\,dy.$$

Let σ_x^2 and σ_y^2 denote the variances of the values of x and of y; then

$$\sigma_x^2=\frac{1}{N}\Sigma(x-\bar{x})^2\,dn$$

$$=\int_{-\infty}^{\infty}\int_{-\infty}^{\infty}(x-\bar{x})^2\,\phi(x,y)\,dx\,dy$$

and
$$\sigma_y^2=\int_{-\infty}^{\infty}\int_{-\infty}^{\infty}(y-\bar{y})^2\,\phi(x,y)\,dx\,dy.$$

The covariance, p, is given by

$$p=\int_{-\infty}^{\infty}\int_{-\infty}^{\infty}(x-\bar{x})\,(y-\bar{y})\,\phi(x,y)\,dx\,dy.$$

The correlation coefficient, r, is defined, as in 9·03 (6), by

$$r=\frac{p}{\sigma_x\sigma_y}.$$

9·09. Normal bivariate distributions

In deriving the normal law for one variable the postulate that the arithmetic mean is the most probable value of the measures was invoked (§ 3·03). In the case of a bivariate distribution the most probable position of the points represented by the coordinates (x_i, y_i) in a scatter diagram, where $i=1, 2, \ldots, n$, is assumed to be the centroid $(\bar{x}, \bar{y})$; this is known as Cotes's postulate.

If ξ_i and η_i are the deviations, from $\bar{x}$ and $\bar{y}$ respectively, of x_i and y_i then

$$\xi_i=x_i-\bar{x}, \quad \eta_i=y_i-\bar{y} \tag{1}$$

and
$$\Sigma\xi_i=0, \quad \Sigma\eta_i=0. \tag{2}$$

Let $\phi(\xi_i, \eta_i)$ denote the probability that a point is associated with deviations ξ_i and η_i. If P denotes the probability that the complete distribution occurs, then

$$P=\prod_{i=1}^{n}\phi(\xi_i, \eta_i)\equiv\prod\phi(x_i-\bar{x}, y_i-\bar{y}).$$

The probability is a maximum when

$$\frac{\partial P}{\partial \bar{x}}=0 \quad \text{and} \quad \frac{\partial P}{\partial \bar{y}}=0.$$

On taking logarithms the first condition is equivalent to

$$\Sigma \frac{\partial}{\partial \bar{x}} \log \phi(x_i-\bar{x}, y_i-\bar{y})=0,$$

or, by (1),
$$\Sigma \frac{\partial}{\partial \xi_i} \log \phi(\xi_i, \eta_i)=0. \tag{3}$$

Similarly, the second condition is equivalent to

$$\Sigma \frac{\partial}{\partial \eta_i} \log \phi(\xi_i, \eta_i)=0.$$

Let $\psi(\xi_i, \eta_i)$ denote $\frac{\partial}{\partial \xi_i} \log \phi(\xi_i, \eta_i)$. Then (3) is

$$\psi(\xi_1, \eta_1)+\psi(\xi_2, \eta_2)+\ldots+\psi(\xi_n, \eta_n)=0. \tag{4}$$

Now, the n ξ's are not independent for, by (2), ξ_n, for example, is given by

$$\xi_n=-(\xi_1+\xi_2+\ldots+\xi_{n-1}). \tag{5}$$

Hence, by differentiation of (4) with respect to ξ_1 we have

$$\frac{\partial \psi(\xi_1, \eta_1)}{\partial \xi_1}+\frac{\partial \psi}{\partial \xi_n} \cdot \frac{\partial \xi_n}{\partial \xi_1}=0,$$

or, by (5),
$$\frac{\partial \psi}{\partial \xi_1}(\xi_1, \eta_1)=\frac{\partial \psi}{\partial \xi_n}(\xi_n, \eta_n).$$

Similar equations are obtained on differentiating with respect to ξ_i, where $i=2, 3, \ldots, n-1$.

Hence, if we write ξ, η simply for any pair ξ_i, η_i we must have

$$\frac{\partial \psi}{\partial \xi}=\text{constant}=-2a.$$

In the same way
$$\frac{\partial \psi}{\partial \eta}=-2h,$$

where h is a constant. Hence

$$d\psi \equiv \frac{\partial \psi}{\partial \xi} d\xi+\frac{\partial \psi}{\partial \eta} d\eta=-2a\, d\xi-2h\, d\eta,$$

whence
$$\psi \equiv \frac{\partial}{\partial \xi} \log \phi(\xi, \eta)=-(2a\xi+2h\eta)+C, \tag{6}$$

where C is a constant. Then, from (4),

$$\Sigma\psi(\xi_i, \eta_i) \equiv -2a\Sigma\xi_i - 2h\Sigma\eta_i + nC = 0,$$

from which, by (2), $C=0$.

By a similar procedure we obtain

$$\frac{\partial}{\partial\eta}\log\phi(\xi,\eta) = -(2h_1\xi + 2b\eta). \tag{7}$$

Differentiate (6) and (7) with respect to η and ξ respectively; then $h_1 = h$.

Now
$$d\log\phi = \frac{\partial(\log\phi)}{\partial\xi}d\xi + \frac{\partial(\log\phi)}{\partial\eta}d\eta;$$

hence
$$\log\phi = -(a\xi^2 + 2h\xi\eta + b\eta^2) + \log k$$

or
$$\phi(\xi,\eta) = k\,e^{-(a\xi^2+2h\xi\eta+b\eta^2)}.$$

It is convenient in the sequel to denote the deviations by x and y; the function ϕ is then given by

$$\phi(x,y) = k\,e^{-(ax^2+2hxy+by^2)}, \tag{8}$$

the origin being at the centroid.

The distribution, given by (8), is a *normal bivariate distribution.*

9·10. The constants of the normal bivariate distribution in terms of σ_x σ_y and r

We now apply the normal function given by 9·09 (8) to a continuous bivariate distribution. The probability that deviations between x and $x+dx$ and between y and $y+dy$ should occur is $\phi(x,y)\,dx\,dy$. The total probability for all such deviations is unity; hence

$$\int_{-\infty}^{\infty}\int_{-\infty}^{\infty}\phi(x,y) = 1,$$

that is,
$$kI = 1, \tag{1}$$
where I is given by

$$I = \int_{-\infty}^{\infty}\int_{-\infty}^{\infty} e^{-(ax^2+2hxy+by^2)}\,dx\,dy. \tag{2}$$

Now,
$$ax^2 + 2hxy + by^2 \equiv \frac{1}{a}(ax+hy)^2 + \frac{1}{a}(ab-h^2)\,y^2.$$

Hence, if
$$\Delta = ab - h^2, \tag{3}$$

$$I = \int_{-\infty}^{\infty}\exp\left[-\frac{\Delta}{a}y^2\right]\left[\int_{-\infty}^{\infty}\exp\left\{-\frac{1}{a}(ax+hy)^2\right\}dx\right]dy.$$

The integral within the squared brackets can be written as

$$\int_{-\infty}^{\infty} e^{-az^2}\,dz, \quad \text{where} \quad z=x+\frac{h}{a}y,$$

and its value for any assigned value of y is $\sqrt{(\pi/a)}$. Thus,

$$I=\sqrt{\left(\frac{\pi}{a}\right)}\int_{-\infty}^{\infty}\exp\left[-\frac{\Delta}{a}y^2\right]dy=\frac{\pi}{\sqrt{\Delta}}, \tag{4}$$

and, from (1),
$$k=\frac{\sqrt{\Delta}}{\pi}.$$

Accordingly,
$$\phi(x,y)=\frac{\sqrt{\Delta}}{\pi}e^{-(ax^2+2hxy+by^2)}. \tag{5}$$

Now,
$$\sigma_x^2=\int_{-\infty}^{\infty}\int_{-\infty}^{\infty}x^2\phi(x,y)\,dx\,dy$$
$$=\frac{\sqrt{\Delta}}{\pi}\int_{-\infty}^{\infty}\int_{-\infty}^{\infty}x^2\,e^{-(ax^2+2hxy+by^2)}\,dx\,dy$$
$$=-\frac{\sqrt{\Delta}}{\pi}\frac{\partial I}{\partial a}.$$

But, from (4) and (3);
$$\frac{\partial I}{\partial a}=-\frac{\pi b}{2\Delta^{\frac{3}{2}}};$$

hence
$$\sigma_x^2=\frac{b}{2\Delta}. \tag{6}$$

Similarly,
$$\sigma_y^2=\frac{a}{2\Delta}. \tag{7}$$

Again,
$$p=\frac{\sqrt{\Delta}}{\pi}\int_{-\infty}^{\infty}\int_{-\infty}^{\infty}xy\,e^{-(ax^2+2hxy+by^2)}\,dx\,dy$$
$$=-\frac{1}{2}\frac{\sqrt{\Delta}}{\pi}\frac{\partial I}{\partial h}.$$

But, from (4) and (3),
$$\frac{\partial I}{\partial h}=-\frac{\pi}{2\Delta^{\frac{3}{2}}}(-2h)=\frac{\pi h}{\Delta^{\frac{3}{2}}}.$$

Hence
$$p=-\frac{h}{2\Delta}. \tag{8}$$

From (6), (7) and (8)
$$\sigma_x^2\sigma_y^2-p^2=\frac{1}{4\Delta},$$

and, since $r = p/(\sigma_x \sigma_y)$, $\sigma_x^2 \sigma_y^2 (1-r^2) = \dfrac{1}{4\Delta}$. (9)

From (7) and (9), we have at once

$$a = \frac{1}{2\sigma_x^2(1-r^2)}. \tag{10}$$

Similarly,

$$b = \frac{1}{2\sigma_y^2(1-r^2)}, \tag{11}$$

and

$$h = -\frac{r}{2\sigma_x \sigma_y (1-r^2)}. \tag{12}$$

Thus, a and b are positive and h is positive or negative.

Also,

$$\Delta = \frac{1}{4\sigma_x^2 \sigma_y^2 (1-r^2)}, \tag{13}$$

so that

$$k = \frac{1}{2\pi \sigma_x \sigma_y \sqrt{(1-r^2)}}. \tag{14}$$

Also, for later use, by means of (6), (7) and (13),

$$\frac{ab}{\Delta} = \frac{1}{1-r^2}. \tag{15}$$

In terms of σ_x, σ_y and r the normal bivariate function is given by

$$\phi(x,y) = \frac{1}{2\pi\sigma_x\sigma_y\sqrt{(1-r^2)}} \exp\left[-\frac{1}{2(1-r^2)}\left\{\frac{x^2}{\sigma_x^2} - \frac{2rxy}{\sigma_x\sigma_y} + \frac{y^2}{\sigma_y^2}\right\}\right].$$

Now,

$$E \equiv ax^2 + 2hxy + by^2 = \frac{\Delta}{b}x^2 + b\left(y + \frac{hx}{b}\right)^2$$

or

$$E = \frac{\Delta}{b}x^2 + b\left(y - \frac{rx\sigma_y}{\sigma_x}\right)^2. \tag{16}$$

Hence

$$\phi(x,y)\,dx\,dy = \left(\frac{\Delta}{\pi b}\right)^{\frac{1}{2}} \exp\left[-\frac{\Delta}{b}x^2\,dx\right]\left(\frac{b}{\pi}\right)^{\frac{1}{2}} \exp\left[-b\left(y - \frac{rx\sigma_y}{\sigma_x}\right)^2\right]dy.$$

This expresses the probability (i) that a point lies in the strip between the lines perpendicular to the x-axis with abscissae x and $x+dx$, together with the probability (ii) that the point lies within this strip with ordinates between y and $y+dy$. The first probability can be written, by means of (6), as

$$\frac{1}{\sigma_x\sqrt{(2\pi)}} e^{-x^2/2\sigma_x^2}\,dx,$$

the function in which is a normal or Gaussian function with standard deviation σ_x. The second probability is normal with respect to the point (x, y_m), where

$$y_m = \frac{r\sigma_y}{\sigma_x} x;$$

the standard deviation in this case is $1/\sqrt{(2b)}$ or $\sigma_y(1-r^2)^{\frac{1}{2}}$. Thus, for a normal bivariate distribution the points (x, y_m) lie on the line of regression of y on x.

By considering a strip between parallels to the x-axis with ordinates y and $y+dy$, we can write

$$\phi(x, y)\, dx\, dy = \left(\frac{\Delta}{\pi a}\right)^{\frac{1}{2}} \exp\left[-\frac{\Delta}{a} y^2\right] dy \left(\frac{a}{\pi}\right)^{\frac{1}{2}} e^{-a(x-x_m)^2}\, dx,$$

where

$$x_m = \frac{r\sigma_x}{\sigma_y} y;$$

the point (x_m, y) lies on the line of regression of x on y.

9·11. The probability ellipse

The probability that deviations from the centroid lie between x and $x+dx$ and between y and $y+dy$ is

$$\phi(x, y)\, dx\, dy = \frac{\sqrt{\Delta}}{\pi} e^{-E}\, dx\, dy, \tag{1}$$

where

$$E = ax^2 + 2hxy + by^2.$$

Now $r^2 < 1$ so that a, b and Δ are positive. Hence, by 9·10 (16), E is positive.

The probability expressed by (1) is constant for all points on the curve $E = \lambda$, where λ is a constant which is necessarily positive. Since Δ is positive the curve is the ellipse

$$ax^2/\lambda + 2hxy/\lambda + by^2/\lambda = 1.$$

Referred to its principal axes the ellipse is

$$x^2/\alpha^2 + y^2/\beta^2 = 1,$$

and we have the invariants for change of axes, namely,

$$\frac{1}{\alpha^2} + \frac{1}{\beta^2} = \frac{1}{\lambda}(a+b)$$

and

$$\frac{1}{\alpha^2\beta^2} = \frac{1}{\lambda^2}(ab - h^2) = \frac{\Delta}{\lambda^2}.$$

The area of the ellipse is $\pi\alpha\beta$ or $\pi\lambda/\Delta^{\frac{1}{2}}$, by the last equation. The area, dA, between the ellipses with parameters λ and $\lambda+d\lambda$ is $\pi d\lambda/\Delta^{\frac{1}{2}}$, and the probability of a point falling within this area is

$$\phi(x,y)\,dA \quad \text{or} \quad \frac{\Delta^{\frac{1}{2}}}{\pi}e^{-\lambda}\frac{\pi\,d\lambda}{\Delta^{\frac{1}{2}}} \quad \text{or} \quad e^{-\lambda}\,d\lambda.$$

Hence the probability of a point falling within the ellipse with parameter λ is

$$\int_0^\lambda e^{-\lambda}\,d\lambda \quad \text{or} \quad 1-e^{-\lambda}.$$

The *probability ellipse* is defined in much the same way as probable error for one variable; it is the ellipse such that the probability of a

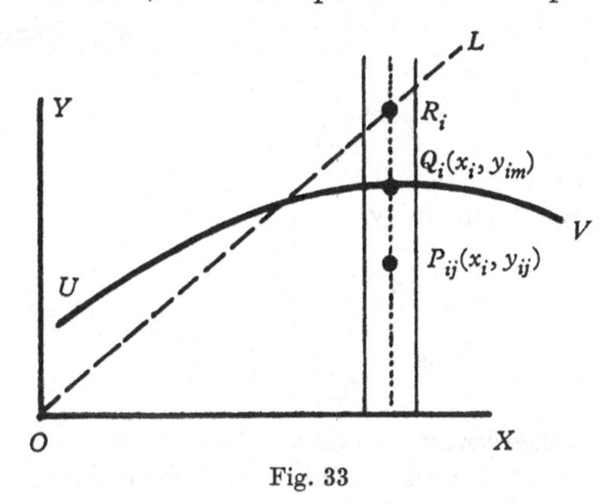

Fig. 33

point falling within it is equal to the probability that the point falls outside it. Hence the parameter for this ellipse is given by

$$1-e^{-\lambda}=\tfrac{1}{2},$$

from which $e^{\lambda}=2$. It is readily found that

$$\lambda=0{\cdot}693.$$

9·12. Curved lines of regression for discrete distributions

Let f_i be the frequency of the points in the vertical strip in Fig. 33 bounded by the ordinates $x=x_i-\frac{1}{2}c$ and $x=x_i+\frac{1}{2}c$, where c is the class interval and *the origin, O, is the centroid*; we suppose that all points in the strip have abscissa x_i. P_{ij} is a representative point in the strip with coordinates (x_i, y_{ij}).

Let y_{im} denote the mean ordinate of points in the strip so that

$$f_i y_{im}=\sum_j y_{ij}. \tag{1}$$

If Q_i is the point corresponding to y_{im}, the locus of Q_i is the *regression curve* of y on x, represented in Fig. 33 by the curve UV.

The correlation will be close for the strip if $\frac{1}{f_i}\Sigma P_{ij}Q_i^2$ is small and for the whole distribution if $\frac{1}{N}\sum_i\sum_j P_{ij}Q_i^2$ is small, where N is the total frequency. Denote this double summation by S^2; then

$$NS^2=\sum_i\sum_j(y_{ij}-y_{im})^2$$

$$=\sum_i\sum_j y_{ij}^2-2\sum_i y_{im}\sum_j y_{ij}+\sum_i(\sum_j y_{im}^2).$$

The first double summation is σ_y^2. Also, $\sum_j y_{im}^2=f_i y_{im}^2$. Hence, by (1),

$$S^2=\sigma_y^2-\frac{1}{N}\sum_i f_i y_{im}^2.$$

For close correlation S is small.

Define η (a positive quantity) by

$$\eta^2=1-\frac{S^2}{\sigma_y^2}; \tag{2}$$

then

$$\eta^2=\frac{1}{N\sigma_y^2}\sum_i f_i y_{im}^2. \tag{3}$$

Here η is called the *correlation ratio* for y on x.

It follows from (2) that η is close to 1 for a high degree of correlation.

For a given bivariate distribution, y_{im} can be found for any vertical array in a contingency table with a suitable class interval; also, σ_y can be obtained in the usual way. Thus, the correlation ratio can be calculated by means of (3). If η proves to be close to 1, the degree of correlation is high.

The correlation ratio for x on y can be found in a similar way.

We now prove two results: (i) if the curve of regression of y on x is a straight line, then $\eta=r$, and (ii) for a curved line of regression, $\eta>r$.

(i) In this case the locus of Q_i is the line

$$y=Mx,$$

where $M=r\sigma_y/\sigma_x$; hence, for Q_i, $y_{im}=Mx_i$ and (3) becomes

$$N\sigma_y^2\eta^2=M^2\sum_i f_i x_i^2=NM^2\sigma_x^2=N\sigma_y^2 r^2.$$

Thus, $\eta=r$.

(ii) Whatever the distributon, the quantities σ_x, σ_y, r and M can be calculated; hence the line $y=Mx$ can be drawn; it is OL in Fig. 33.

If the ordinate through Q_i meets OL at R_i, then Q_iR_i measures the departure of the curved regression line from OL for the strip. Now

$$Q_iR_i^2 = (y_{im} - Mx_i)^2$$

and

$$\sum_i f_i \cdot Q_iR_i^2 = \Sigma f_i y_{im}^2 - 2M\,\Sigma f_i x_i y_{im} + M^2 \Sigma f_i x_i^2$$
$$= N\sigma_y^2\eta^2 - 2MNp + M^2N\sigma_x^2, \qquad (4)$$

where p is the covariance which is equal to $r\sigma_x\sigma_y$. Hence, with $M = r\sigma_y/\sigma_x$, the right-hand side of (4) reduces to $N\sigma_y^2(\eta^2 - r^2)$. Since the left-hand side of (4) is positive, then $\eta > r$.

In this case, when the line of regression is not a straight line, the correlation ratio, η, is to be preferred as a suitable measure of correlation over the correlation coefficient, r.

9·13. Correlation ratio for a continuous bivariate distribution

In Fig. 33 let the ordinates shown define a strip of width dx. If the probability function is $\phi(x, y)$ with the property

$$\int_{-\infty}^{\infty}\int_{-\infty}^{\infty} \phi(x, y)\,dx\,dy = 1,$$

and if N, a large number, is the total frequency, the frequency for the element $dx\,dy$ is $N\phi(x, y)\,dx\,dy$.

Let $N_x dx$ denote the frequency for the whole strip; then

$$N_x dx = N\,dx \int_{-\infty}^{\infty} \phi(x, y)\,dy. \qquad (1)$$

The mean, y_m, of the ordinates in the strip is given by

$$N_x dx\, y_m = N dx \int_{-\infty}^{\infty} y\phi(x, y)\,dy.$$

Now, y_m is a function of x and we can write this last equation as

$$N_x dx\, y_m^2 = N\,dx \int_{-\infty}^{\infty} y y_m \phi(x, y)\,dy,$$

or, by means of (1) and on integration with respect to x,

$$\int_{-\infty}^{\infty}\int_{-\infty}^{\infty} y_m^2 \phi(x, y)\,dx\,dy = \int_{-\infty}^{\infty}\int_{-\infty}^{\infty} y y_m \phi(x, y)\,dx\,dy. \qquad (2)$$

Let s_y^2 be the variance for the strip with reference to the mean y_m; then

$$N_x dx\, s_y^2 = N\,dx \int_{-\infty}^{\infty} (y - y_m)^2\,\phi(x, y)\,dy. \qquad (3)$$

So far as the strip is concerned, the correlation will be high if s_y is small.

Let NS^2 denote $\int_{-\infty}^{\infty} N_x s_y^2 dx$; then, from (3),

$$S^2 = \int_{-\infty}^{\infty}\int_{-\infty}^{\infty} (y-y_m)^2 \phi(x,y)\,dx\,dy,$$

which can be written as

$$S^2 = \int_{-\infty}^{\infty}\int_{-\infty}^{\infty} y^2\phi(x,y)\,dx\,dy - \int_{-\infty}^{\infty}\int_{-\infty}^{\infty} y_m^2\phi(x,y)\,dx\,dy$$
$$-2\int_{-\infty}^{\infty}\int_{-\infty}^{\infty} y_m(y-y_m)\,\phi(x,y)\,dx\,dy.$$

The first double integral is σ_y^2 and the last double integral vanishes by (2). Hence

$$S^2 = \sigma_y^2 - \int_{-\infty}^{\infty}\int_{-\infty}^{\infty} y_m^2\,\phi(x,y)\,dx\,dy.$$

Defining the correlation ratio, as before, by

$$\eta^2 = 1 - \frac{S^2}{\sigma_y^2},$$

we obtain
$$\eta^2 = \frac{1}{\sigma_y^2}\int_{-\infty}^{\infty}\int_{-\infty}^{\infty} y_m^2\,\phi(x,y)\,dx\,dy.$$

The correlation will be close if η is near 1. It follows as in the previous section that if the distribution is normal then $\eta = r$ and that, in all other cases, $\eta > r$.

9·14. Contingency coefficient

In the previous sections on bivariate distributions the attributes have, by implication, been capable of numerical expression, as illustrated in Table 21, which gives the statistics relating to the measures of stellar radii and masses. In some problems one attribute, or even both, may signify a quality which cannot be specified numerically: for example, we may be dealing with groups of stars, each group being associated with a particular colour representing one attribute.

Let A_i $(i = 1, 2, \ldots, m)$ represent the various groups arranged according to one particular quality and B_j $(j = 1, 2, \ldots, n)$, the groups arranged according to a second quality, or perhaps to a numerical specification. We can then form the contingency table represented in outline by Table 24, in which n_{ij} denotes the frequency of individuals with attributes A_i and B_j, a_i denotes the sum of all the individuals with the attribute A_i, and b_j denotes the sum of all the individuals with the attribute B_j. If N is the total frequency, then

$$N = \sum_i a_i = \sum_j b_j.$$

Table 24

	A_1	A_2	$\dots$	A_i	$\dots$	A_m	Sums
B_1	n_{11}	n_{11}	$\dots$	$\dots$	$\dots$	n_{m1}	b_1
B_2	n_{12}	$\dots$	$\dots$	$\dots$	$\dots$	$\dots$	b_2
$\vdots$							$\vdots$
B_j	$\dots$	$\dots$	$\dots$	n_{ij}	$\dots$	$\dots$	b_j
$\vdots$							$\vdots$
B_n	n_{1n}	$\dots$	$\dots$	$\dots$	$\dots$	n_{mn}	b_n
Sums	a_1	a_2	$\dots$	a_i	$\dots$	a_m	

From the data we infer that the probability that an individual will have the property A_i is a_i/N and, similarly, that the probability that an individual will have the property B_j is b_j/N. For *unrelated* attributes let ν_{ij}/N denote the probability that an individual will have the property A_i *and* the property B_j; then

$$\nu_{ij}/N = \frac{a_i}{N} \cdot \frac{b_j}{N}$$

or

$$\nu_{ij} = \frac{1}{N} a_i b_j.$$

We can now form, from the statistics, the contingency Table 25 in terms of the numbers ν_{ij}.

Table 25

	A_1	$\dots$	A_i	$\dots$	A_m
B_1	ν_{11}	$\dots$	ν_{i1}	$\dots$	ν_{m1}
$\vdots$					
B_j	$\dots$	$\dots$	ν_{ij}	$\dots$	$\dots$
$\vdots$					
B_n	ν_{1n}	$\dots$	$\dots$	$\dots$	ν_{mn}

If the attributes are associated by pure chance, the statistics within the Tables 24 and 25 may be expected to be very much alike for ordinary values of N and identical if N is a very large number.

If, on the other hand, there is some degree of correlation, the differences $(n_{ij} - \nu_{ij})$ may be expected to be systematic in character.

Pearson's measure of correlation is defined in terms of a *contingency function*,† ψ, given by

$$\psi^2 = \frac{1}{N} \sum_i \sum_j \frac{(n_{ij} - \nu_{ij})^2}{\nu_{ij}},$$

† The function introduced by Pearson is usually denoted by ϕ but, to obviate confusion with the normal bivariate function, ϕ, of the previous sections, Pearson's function is denoted here by ψ.

and the *contingency coefficient*, c, related to the degree of correlation, is defined by

$$c = \frac{\psi}{\sqrt{(1+\psi^2)}},$$

in which $0 < c < 1$.

Thus, when the association of the attributes is one of pure chance and N is very large, then $\psi = 0$ and, accordingly, $c = 0$. When the correlation is considerable, then c will have a high fractional value.

9·15. The contingency function ψ for a normal bivariate distribution

For a normal bivariate distribution the contingency function can be evaluated in terms of the correlation coefficient, r. The normal bivariate function, ϕ, is given, from 9·10 (5), by

$$\phi(x, y) = \frac{\Delta^{\frac{1}{2}}}{\pi} e^{-E},$$

where $E \equiv ax^2 + 2hxy + by^2$; from the property

$$\int_{-\infty}^{\infty}\int_{-\infty}^{\infty} \phi(x, y)\, dx\, dy = 1,$$

we have

$$\int_{-\infty}^{\infty}\int_{-\infty}^{\infty} e^{-E}\, dx\, dy = \frac{\pi}{\Delta^{\frac{1}{2}}}. \tag{1}$$

The contingency table for the statistics may now be supposed to have a typical class interval dx (corresponding to A in Table 24) and a class interval dy (corresponding to B).

Then, if $a_i dx$ denotes the frequency in the interval x_i to $x_i + dx$ for all values of y, the total frequency being N, we have

$$a_i dx = N\, dx \int_{-\infty}^{\infty} \phi(x, y)\, dy = N \frac{\Delta^{\frac{1}{2}}}{\pi} dx \int_{-\infty}^{\infty} e^{-E}\, dy.$$

Since

$$E \equiv \frac{\Delta}{b} x^2 + \frac{(by + hx)^2}{b},$$

we obtain, in the usual way,

$$a_i dx = \frac{N\Delta^{\frac{1}{2}}}{\sqrt{(\pi b)}} \exp\left[-\frac{\Delta}{b} x^2\right] dx.$$

Similarly,

$$b_j dy = \frac{N\Delta^{\frac{1}{2}}}{\sqrt{(\pi a)}} \exp\left[-\frac{\Delta}{a} y^2\right] dy.$$

Accordingly,

$$\nu_{ij}\, dx\, dy \equiv \frac{1}{N} a_i dx\, b_j dy = \frac{N\Delta}{\pi\sqrt{(ab)}} \exp\left[-\frac{\Delta}{b} x^2 - \frac{\Delta}{a} y^2\right].$$

Also,
$$n_{ij}\,dx\,dy=\frac{N\Delta^{\frac{1}{2}}}{\pi}e^{-E}\,dx\,dy.$$

From these we have

$$\int_{-\infty}^{\infty}\int_{-\infty}^{\infty}\nu_{ij}\,dx\,dy=N=\int_{-\infty}^{\infty}\int_{-\infty}^{\infty}n_{ij}\,dx\,dy \tag{2}$$

and
$$\frac{n_{ij}^2}{\nu_{ij}}=\frac{N\sqrt{(ab)}}{\pi}e^{-E_1}, \tag{3}$$

where
$$E_1\equiv 2E-\frac{\Delta}{b}x^2-\frac{\Delta}{a}y^2$$
$$\equiv a_1x^2+2h_1xy+b_1y^2,$$

in which
$$a_1=2a-\frac{\Delta}{b}=\frac{ab+h^2}{b},$$
$$h_1=2h,$$
$$b_1=\frac{ab+h^2}{a}.$$

Then,
$$\Delta_1\equiv a_1b_1-h_1^2=\frac{(ab-h^2)^2}{ab}=\frac{\Delta^2}{ab}.$$

By analogy with (1),

$$\int_{-\infty}^{\infty}\int_{-\infty}^{\infty}e^{-E_1}\,dx\,dy=\frac{\pi}{\Delta_1^{\frac{1}{2}}}=\frac{\pi\sqrt{(ab)}}{\Delta};$$

hence, from (3),
$$\int_{-\infty}^{\infty}\int_{-\infty}^{\infty}\frac{n_{ij}^2}{\nu_{ij}}\,dx\,dy=\frac{Nab}{\Delta}$$
$$=\frac{N}{1-r^2},$$

by means of 9·10 (15).

Now the contingency function, ψ, is given in this case by

$$N\psi^2\equiv\int_{-\infty}^{\infty}\int_{-\infty}^{\infty}\frac{(n_{ij}-\nu_{ij})^2}{\nu_{ij}}\,dx\,dy$$
$$=\frac{N}{1-r^2}-2\int_{-\infty}^{\infty}\int_{-\infty}^{\infty}n_{ij}\,dx\,dy+\int_{-\infty}^{\infty}\int_{-\infty}^{\infty}\nu_{ij}\,dx\,dy$$
$$=N\left(\frac{1}{1-r^2}-1\right),\quad\text{by (2).}$$

Hence
$$\psi^2=\frac{r^2}{1-r^2}.$$

The contingency coefficient, c, is then given by

$$c \equiv \frac{\psi}{\sqrt{(1+\psi^2)}} = r.$$

Thus, for a normal bivariate distribution, the contingency coefficient is identical with the correlation coefficient, r.

9·16. Normal trivariate distributions

As in the case of a bivariate distribution, the most probable position of the points represented by the coordinates (x_i, y_i, z_i) is assumed to be the centroid $(\bar{x}, \bar{y}, \bar{z})$.

If ξ_i, η_i, ζ_i are the deviations from $\bar{x}, \bar{y}, \bar{z}$, respectively, of x_i, y_i, z_i then

$$\xi_i = x_i - \bar{x}, \quad \eta_i = y_i - \bar{y}, \quad \zeta_i = z_i - \bar{z}.$$

The relative frequency function referred to axes through the centroid is denoted by

$$\phi(\xi_i, \eta_i, \zeta_i) \equiv \phi(x_i - \bar{x}, y_i - \bar{y}, z_i - \bar{z}).$$

The probability, P, for the distribution is given by

$$P = \Pi \phi(x_i - \bar{x}, y_i - \bar{y}, z_i - \bar{z}),$$

and P is a maximum when

$$\frac{\partial P}{\partial \bar{x}} = \frac{\partial P}{\partial \bar{y}} = \frac{\partial P}{\partial \bar{z}} = 0.$$

From the first condition we have, on taking logarithms,

$$\Sigma \frac{\partial}{\partial \bar{x}} \log \phi \equiv \Sigma \frac{\partial}{\partial \xi_i} \log \phi(\xi_i, \eta_i, \zeta_i) = 0.$$

But $\Sigma \xi_i = 0$; hence, by arguments similar to those leading to 9·09 (6), we obtain, on writing (ξ, η, ζ) for a representative point,

$$\frac{\partial}{\partial \xi} \log \phi(\xi, \eta, \zeta) = -(2a\xi + 2h\eta + 2g\zeta).$$

It is convenient, as before, to denote the deviations ξ, η, ζ by x, y, z; then

$$\frac{\partial}{\partial x} \log \phi(x, y, z) = -(2ax + 2hy + 2gz).$$

The second and third conditions in (1) give, similarly,

$$\frac{\partial}{\partial y} \log \phi = -(2hx + 2by + 2fz)$$

and $$\frac{\partial}{\partial z}\log\phi = -(2gx+2fy+2cz),$$

which are seen to satisfy the necessary conditions

$$\frac{\partial}{\partial x}\frac{\partial}{\partial y} = \frac{\partial}{\partial y}\frac{\partial}{\partial x}, \quad \text{etc.}$$

We then obtain

$$d(\log\phi) \equiv \frac{\partial}{\partial x}\log\phi\,dx + \frac{\partial}{\partial y}\log\phi\,dy + \frac{\partial}{\partial z}\log\phi\,dz$$
$$= -d(ax^2+by^2+cz^2+2fyz+2gzx+2hxy);$$

hence $$\phi(xyz) = k\,e^{-E}, \tag{2}$$

where $$E \equiv ax^2+by^2+cz^2+2fyz+2gzx+2hxy, \tag{3}$$

and k is a constant with a specific value which will be found in the next section.

9·17. The constants of the normal trivariate distribution

The probability function $\phi(x, y, z)$ is such that

$$\iiint_{-\infty}^{\infty} \phi\,dx\,dy\,dz = 1,$$

that is, from 9·16 (2), $$kI = 1 \tag{1}$$

where $$I = \iiint_{-\infty}^{\infty} e^{-E}\,dx\,dy\,dz. \tag{2}$$

Now, by 9·16 (3),

$$E \equiv ax^2+2hxy+by^2+c\left(z+\frac{fy+gx}{c}\right)^2 - \frac{(fy+gx)^2}{c}$$
$$= \frac{1}{c}[(ac-g^2)\,x^2+2(ch-fg)\,xy+(bc-f^2)\,y^2]+c(z+p)^2,$$

where $p = (fy+gx)/c$.

Let Δ be the determinant defined by

$$\Delta = \begin{vmatrix} a, & h, & g \\ h, & b, & f \\ g, & f, & c \end{vmatrix},$$

for which, in the usual notation,

$$A = bc-f^2, \qquad B = ca-g^2, \qquad C = ab-h^2,$$
$$F = gh-af, \qquad G = hf-bg, \qquad H = fg-ch.$$

16-2

Then, for specific values of x and y, or for a specific value of p, we have, from (2),

$$I=\int_{-\infty}^{\infty}\int_{-\infty}^{\infty} e^{-E_1}\,dx\,dy\int_{-\infty}^{\infty} e^{-c(z+p)^2}\,dz,$$

where $$E_1=\frac{1}{c}(Bx^2-2Hxy+Ay^2).$$

The z-integral is $\sqrt{(\pi/c)}$.

Further, $$E_1=\frac{A}{c}\left(y-\frac{H}{A}x\right)^2+\frac{1}{cA}(AB-H^2)\,x^2.$$

But $$AB-H^2=c\Delta;$$

hence

$$I=\sqrt{\left(\frac{\pi}{c}\right)}\int_{-\infty}^{\infty}\exp\left[-\frac{\Delta}{A}x^2\right]dx\int_{-\infty}^{\infty}\exp\left[-\frac{A}{c}\left(y-\frac{H}{A}x\right)^2\right]dy.$$

so that, finally, $$I=\frac{\pi^{\frac{3}{2}}}{\Delta^{\frac{1}{2}}}. \tag{3}$$

Then, from (1), $$k=\frac{\Delta^{\frac{1}{2}}}{\pi^{\frac{3}{2}}}.$$

The variance, σ_x^2, for the values of x in the distribution is given by

$$\sigma_x^2=\iiint_{-\infty}^{\infty} x^2\phi(x,y,z)\,dx\,dy\,dz$$

$$=k\iiint_{-\infty}^{\infty} x^2\,e^{-E}\,dx\,dy\,dz$$

$$=-k\frac{\partial I}{\partial a}.$$

Now, from (3), $$\frac{\partial I}{\partial a}=-\frac{\pi^{\frac{3}{2}}}{2\Delta^{\frac{3}{2}}}\frac{\partial\Delta}{\partial a}=-\frac{1}{2k\Delta}\frac{\partial\Delta}{\partial a}.$$

But, $$\Delta=abc+2fgh-af^2-bg^2-ch^2,$$

and hence $$\frac{\partial\Delta}{\partial a}=bc-f^2=A,$$

and $$\sigma_x^2=\frac{A}{2\Delta}.$$

Similarly, $$\sigma_y^2=\frac{B}{2\Delta},\quad \sigma_z^2=\frac{C}{2\Delta}.$$

Again, the covariance, p_{xy}, for the variables x and y is given by

$$p_{xy}=\iiint\limits_{-\infty}^{\infty} xy\phi(x,y,z)\,dx\,dy\,dz=k\iiint\limits_{-\infty}^{\infty} xy\,e^{-E}\,dx\,dy\,dz$$

$$=-\tfrac{1}{2}k\frac{\partial I}{\partial h}.$$

But $$\frac{\partial I}{\partial h}=-\frac{\pi^{\frac{3}{2}}}{2\Delta^{\frac{3}{2}}}\frac{\partial\Delta}{\partial h}=-\frac{H}{k\Delta}.$$

Hence $$p_{xy}=\frac{H}{2\Delta}.$$

Similarly, $$p_{yz}=\frac{F}{2\Delta},\quad p_{zx}=\frac{G}{2\Delta}.$$

The correlation coefficient for the variables x and y, denoted by r_{xy}, is defined by

$$r_{xy}=\frac{p_{xy}}{\sigma_x\sigma_y};$$

hence $$r_{xy}\sigma_x\sigma_y=\frac{H}{2\Delta}.$$

Similarly, $$r_{yz}\sigma_y\sigma_z=\frac{F}{2\Delta},\quad r_{zx}\sigma_z\sigma_x=\frac{G}{2\Delta}.$$

Now,

$$\Delta^2=\begin{vmatrix} A, & H, & G \\ H, & B, & F \\ G, & F, & C \end{vmatrix}=8\Delta^3\begin{vmatrix} \sigma_x^2, & r_{xy}\sigma_x\sigma_y, & r_{xz}\sigma_x\sigma_z \\ r_{xy}\sigma_x\sigma_y, & \sigma_y^2, & r_{yz}\sigma_y\sigma_z \\ r_{xz}\sigma_x\sigma_z, & r_{yz}\sigma_y\sigma_z, & \sigma_z^2 \end{vmatrix},$$

from which $$\frac{1}{8\Delta}=\sigma_x^2\sigma_y^2\sigma_z^2\begin{vmatrix} 1, & r_{xy}, & r_{xz} \\ r_{xy}, & 1, & r_{yz} \\ r_{xz}, & r_{yz}, & 1 \end{vmatrix}. \qquad (4)$$

From the statistics σ_x, σ_y and σ_z are found in the usual way; also, the correlation coefficients are each found as illustrated in 9·07. Then Δ is calculated from (4).

Now, $$a\Delta=BC-F^2;$$

hence $$a=4\Delta\sigma_y^2\sigma_z^2(1-r_{yz}^2),$$

with two similar formulae for b and c.

Again, $$f\Delta=GH-AF;$$

hence $$f=4\Delta\sigma_x^2\sigma_y\sigma_z(r_{xy}r_{xz}-r_{yz}),$$

with two similar formulae for g and h.

Thus, the constants in the homogeneous quadratic function E are determined.

APPENDIX 1

VALUES OF $\operatorname{erf} t = \frac{2}{\sqrt{\pi}} \int_0^t e^{-t^2}\, dt$

t	erf t	*Diff.*	t	erf t	*Diff.*	t	erf t	*Diff.*
0·000	0·00000	*564*	**0·250**	0·27633	*529*	**0·500**	0·52050	*438*
0·005	0·00564	*564*	**0·255**	0·28162	*528*	**0·505**	0·52488	*436*
0·010	0·01128	*564*	**0·260**	0·28690	*527*	**0·510**	0·52924	*434*
0·015	0·01692	*564*	**0·265**	0·29217	*525*	**0·515**	0·53358	*432*
0·020	0·02256	*564*	**0·270**	0·29742	*524*	**0·520**	0·53790	*429*
0·025	0·02820	*564*	**0·275**	0·30266	*522*	**0·525**	0·54219	*427*
0·030	0·03384	*564*	**0·280**	0·30788	*521*	**0·530**	0·54646	*425*
0·035	0·03948	*563*	**0·285**	0·31309	*519*	**0·535**	0·55071	*423*
0·040	0·04511	*563*	**0·290**	0·31828	*518*	**0·540**	0·55494	*420*
0·045	0·05074	*563*	**0·295**	0·32346	*517*	**0·545**	0·55914	*418*
0·050	0·05637	*563*	**0·300**	0·32863	*515*	**0·550**	0·56332	*416*
0·055	0·06200	*562*	**0·305**	0·33378	*513*	**0·555**	0·56748	*414*
0·060	0·06762	*562*	**0·310**	0·33891	*512*	**0·560**	0·57162	*411*
0·065	0·07324	*562*	**0·315**	0·34403	*510*	**0·565**	0·57573	*409*
0·070	0·07886	*561*	**0·320**	0·34913	*508*	**0·570**	0·57982	*406*
0·075	0·08447	*561*	**0·325**	0·35421	*507*	**0·575**	0·58388	*404*
0·080	0·09008	*560*	**0·330**	0·35928	*505*	**0·580**	0·58792	*402*
0·085	0·09568	*560*	**0·335**	0·36433	*503*	**0·585**	0·59194	*400*
0·090	0·10128	*559*	**0·340**	0·36936	*502*	**0·590**	0·59594	*397*
0·095	0·10687	*559*	**0·345**	0·37438	*500*	**0·595**	0·59991	*395*
0·100	0·11246	*558*	**0·350**	0·37938	*498*	**0·600**	0·60386	*392*
0·105	0·11804	*558*	**0·355**	0·38436	*497*	**0·605**	0·60778	*390*
0·110	0·12362	*557*	**0·360**	0·38933	*495*	**0·610**	0·61168	*388*
0·115	0·12919	*557*	**0·365**	0·39428	*493*	**0·615**	0·61556	*385*
0·120	0·13476	*556*	**0·370**	0·39921	*491*	**0·620**	0·61941	*383*
0·125	0·14032	*555*	**0·375**	0·40412	*489*	**0·625**	0·62324	*381*
0·130	0·14587	*554*	**0·380**	0·40901	*487*	**0·630**	0·62705	*378*
0·135	0·15141	*554*	**0·385**	0·41388	*486*	**0·635**	0·63083	*376*
0·140	0·15695	*553*	**0·390**	0·41874	*484*	**0·640**	0·63459	*373*
0·145	0·16248	*552*	**0·395**	0·42358	*481*	**0·645**	0·63832	*371*
0·150	0·16800	*551*	**0·400**	0·42839	*480*	**0·650**	0·64203	*369*
0·155	0·17351	*550*	**0·405**	0·43319	*478*	**0·655**	0·64572	*366*
0·160	0·17901	*550*	**0·410**	0·43797	*476*	**0·660**	0·64938	*363*
0·165	0·18451	*548*	**0·415**	0·44273	*474*	**0·665**	0·65301	*362*
0·170	0·18999	*548*	**0·420**	0·44747	*472*	**0·670**	0·65663	*359*
0·175	0·19547	*547*	**0·425**	0·45219	*470*	**0·675**	0·66022	*356*
0·180	0·20094	*545*	**0·430**	0·45689	*468*	**0·680**	0·66378	*354*
0·185	0·20639	*545*	**0·435**	0·46157	*466*	**0·685**	0·66732	*352*
0·190	0·21184	*544*	**0·440**	0·46623	*463*	**0·690**	0·67084	*349*
0·195	0·21728	*542*	**0·445**	0·47086	*462*	**0·695**	0·67433	*347*
0·200	0·22270	*542*	**0·450**	0·47548	*460*	**0·700**	0·67780	*345*
0·205	0·22812	*540*	**0·455**	0·48008	*458*	**0·705**	0·68125	*342*
0·210	0·23352	*539*	**0·460**	0·48466	*455*	**0·710**	0·68467	*339*
0·215	0·23891	*539*	**0·465**	0·48921	*454*	**0·715**	0·68806	*337*
0·220	0·24430	*537*	**0·470**	0·49375	*451*	**0·720**	0·69143	*335*
0·225	0·24967	*535*	**0·475**	0·49826	*449*	**0·725**	0·69478	*332*
0·230	0·25502	*535*	**0·480**	0·50275	*447*	**0·730**	0·69810	*330*
0·235	0·26037	*533*	**0·485**	0·50722	*445*	**0·735**	0·70140	*328*
0·240	0·26570	*532*	**0·490**	0·51167	*443*	**0·740**	0·70468	*325*
0·245	0·27102	*531*	**0·495**	0·51610	*440*	**0·745**	0·70793	*323*
0·250	0·27633		**0·500**	0·52050		**0·750**	0·71116	

VALUES OF erf t (*continued*)

t	erf t	*Diff.*	t	erf t	*Diff.*	t	erf t	*Diff.*
0·75	0·71116	*638*	**1·25**	0·92290	*234*	**1·75**	0·98667	*52*
0·76	0·71754	*628*	**1·26**	0·92524	*227*	**1·76**	0·98719	*50*
0·77	0·72382	*619*	**1·27**	0·92751	*222*	**1·77**	0·98769	*48*
0·78	0·73001	*609*	**1·28**	0·92973	*217*	**1·78**	0·98817	*47*
0·79	0·73610	*600*	**1·29**	0·93190	*211*	**1·79**	0·98864	*45*
0·80	0·74210	*590*	**1·30**	0·93401	*205*	**1·80**	0·98909	*43*
0·81	0·74800	*581*	**1·31**	0·93606	*201*	**1·81**	0·98952	*42*
0·82	0·75381	*571*	**1·32**	0·93807	*195*	**1·82**	0·98994	*41*
0·83	0·75952	*562*	**1·33**	0·94002	*189*	**1·83**	0·99035	*39*
0·84	0·76514	*553*	**1·34**	0·94191	*185*	**1·84**	0·99074	*37*
0·85	0·77067	*543*	**1·35**	0·94376	*180*	**1·85**	0·99111	*36*
0·86	0·77610	*534*	**1·36**	0·94556	*175*	**1·86**	0·99147	*35*
0·87	0·78144	*525*	**1·37**	0·94731	*171*	**1·87**	0·99182	*34*
0·88	0·78669	*515*	**1·38**	0·94902	*165*	**1·88**	0·99216	*32*
0·89	0·79184	*507*	**1·39**	0·95067	*162*	**1·89**	0·99248	*31*
0·90	0·79691	*497*	**1·40**	0·95229	*157*	**1·90**	0·99279	*30*
0·91	0·80188	*489*	**1·41**	0·95386	*152*	**1·91**	0·99309	*29*
0·92	0·80677	*479*	**1·42**	0·95538	*148*	**1·92**	0·99338	*28*
0·93	0·81156	*471*	**1·43**	0·95686	*144*	**1·93**	0·99366	*26*
0·94	0·81627	*462*	**1·44**	0·95830	*140*	**1·94**	0·99392	*26*
0·95	0·82089	*453*	**1·45**	0·95970	*135*	**1·95**	0·99418	*25*
0·96	0·82542	*445*	**1·46**	0·96105	*132*	**1·96**	0·99443	*23*
0·97	0·82987	*436*	**1·47**	0·96237	*128*	**1·97**	0·99466	*23*
0·98	0·83423	*428*	**1·48**	0·96365	*125*	**1·98**	0·99489	*22*
0·99	0·83851	*419*	**1·49**	0·96490	*121*	**1·99**	0·99511	*21*
1·00	0·84270	*411*	**1·50**	0·96611	*117*	**2·00**	0·99532	*20*
1·01	0·84681	*403*	**1·51**	0·96728	*113*	**2·01**	0·99552	*20*
1·02	0·85084	*394*	**1·52**	0·96841	*111*	**2·02**	0·99572	*19*
1·03	0·85478	*387*	**1·53**	0·96952	*107*	**2·03**	0·99591	*18*
1·04	0·85865	*379*	**1·54**	0·97059	*103*	**2·04**	0·99609	*17*
1·05	0·86244	*370*	**1·55**	0·97162	*101*	**2·05**	0·99626	*16*
1·06	0·86614	*363*	**1·56**	0·97263	*97*	**2·06**	0·99642	*16*
1·07	0·86977	*356*	**1·57**	0·97360	*95*	**2·07**	0·99658	*15*
1·08	0·87333	*347*	**1·58**	0·97455	*91*	**2·08**	0·99673	*15*
1·09	0·87680	*341*	**1·59**	0·97546	*89*	**2·09**	0·99688	*14*
1·10	0·88021	*332*	**1·60**	0·97635	*86*	**2·10**	0·99702	*13*
1·11	0·88353	*326*	**1·61**	0·97721	*83*	**2·11**	0·99715	*13*
1·12	0·88679	*318*	**1·62**	0·97804	*80*	**2·12**	0·99728	*13*
1·13	0·88997	*311*	**1·63**	0·97884	*78*	**2·13**	0·99741	*12*
1·14	0·89308	*304*	**1·64**	0·97962	*76*	**2·14**	0·99753	*11*
1·15	0·89612	*298*	**1·65**	0·98038	*72*	**2·15**	0·99764	*11*
1·16	0·89910	*290*	**1·66**	0·98110	*71*	**2·16**	0·99775	*10*
1·17	0·90200	*284*	**1·67**	0·98181	*68*	**2·17**	0·99785	*10*
1·18	0·90484	*277*	**1·68**	0·98249	*66*	**2·18**	0·99795	*10*
1·19	0·90761	*270*	**1·69**	0·98315	*64*	**2·19**	0·99805	*9*
1·20	0·91031	*265*	**1·70**	0·98379	*62*	**2·20**	0·99814	*8*
1·21	0·91296	*257*	**1·71**	0·98441	*59*	**2·21**	0·99822	*9*
1·22	0·91553	*252*	**1·72**	0·98500	*58*	**2·22**	0·99831	*8*
1·23	0·91805	*246*	**1·73**	0·98558	*55*	**2·23**	0·99839	*7*
1·24	0·92051	*239*	**1·74**	0·98613	*54*	**2·24**	0·99846	*8*
1·25	0·92290		**1·75**	0·98667		**2·25**	0·99854	

APPENDIX 1(*a*)

VALUES OF (1 – erf *t*)

t	$1-\text{erf}\,t$	t	$1-\text{erf}\,t$	t	$1-\text{erf}\,t$
2·25	$10^{-6}.1463$	2·55	$10^{-7}.3107$	2·85	$10^{-8}.5566$
2·26	1393	2·56	2942	2·90	4110
2·27	1326	2·57	2785	2·95	3020
2·28	1262	2·58	2636	3·0	2209
2·29	1201	2·59	2495	3·1	$10^{-8}.1165$
2·30	1143	2·60	2360	3·2	$10^{-9}.6026$
2·31	1088	2·61	2233	3·3	3058
2·32	$10^{-6}.1034$	2·62	2112	3·4	$10^{-9}.1522$
2·33	$10^{-7}.9838$	2·63	1997	3·5	$10^{-10}.7411$
2·34	9354	2·64	1888	3·6	3559
2·35	8893	2·65	1785	3·7	$10^{-10}.1672$
2·36	8452	2·66	1687	3·8	$10^{-11}.7700$
2·37	8032	2·67	1594	3·9	3479
2·38	7631	2·68	1506	4·0	$10^{-11}.1522$
2·39	7249	2·69	1422	4·1	$10^{-12}.7000$
2·40	6885	2·70	1343	4·2	2855
2·41	6538	2·71	1268	4·3	1193
2·42	6207	2·72	1197	4·4	489
2·43	5892	2·73	1130	4·5	$10^{-12}.\ 197$
2·44	5592	2·74	1066	4·6	$10^{-13}.\ 775$
2·45	5306	2·75	$10^{-7}.1006$	4·7	200
2·46	5034	2·76	$10^{-8}.9492$	4·8	$10^{-13}.\ 114$
2·47	4774	2·77	8952	4·9	$10^{-14}.\ 422$
2·48	4528	2·78	8441	5·0	$10^{-14}.\ 154$
2·49	4293	2·79	7958	5·5	$10^{-15}.\ \ 7$
2·50	4070	2·80	7501	∞	0
2·51	3857	2·81	7069		
2·52	3655	2·82	6661		
2·53	3463	2·83	6275		
2·54	3280	2·84	5910		
2·55	$10^{-7}.3107$	2·85	$10^{-8}.5566$		

APPENDIX 2

VALUES OF $G(q)$

q	$G(q)$	q	$G(q)$	q	$G(q)$
0·0	0·8862	9·0	3·0846	30·0	5·55
1·0	1·2819	10·0	3·2424	40·0	6·38
2·0	1·6068	11·0	3·3929	50·0	7·13
3·0	1·8849	12·0	3·5371	60·0	7·80
4·0	2·1302	13·0	3·6756	70·0	8·41
5·0	2·3513	14·0	3·8091	80·0	8·98
6·0	2·5541	15·0	3·9381	90·0	9·51
7·0	2·7422	16·0	4·0630	100·0	10·02
8·0	2·9183	20·0	4·52	150·0	12·31
9·0	3·0846	30·0	5·55	—	—

APPENDIX 3

VALUES OF $e^{-k^2x^2}$

x \ k	0·1	0·2	0·3	0·4	0·5	0·6
0·0	1·000	1·000	1·000	1·000	1·000	1·000
0·5	0·998	0·990	0·978	0·961	0·939	0·914
1·0	0·990	0·961	0·914	0·852	0·779	0·698
1·5	0·978	0·914	0·817	0·698	0·570	0·445
2·0	0·961	0·852	0·698	0·527	0·368	0·237
2·5	0·939	0·779	0·570	0·368	0·210	0·105
3·0	0·914	0·698	0·445	0·237	0·105	0·039
3·5	0·885	0·613	0·332	0·141	0·047	0·012
4·0	0·852	0·527	0·237	0·077	0·018	0·003
5·0	0·779	0·368	0·105	0·018	0·002	—
6·0	0·698	0·237	0·039	0·003	—	—
7·0	0·613	0·141	0·012	0·000	—	—
8·0	0·527	0·077	0·003	—	—	—

INDEX

[*The references are to pages*]